ESSENTIAL PRACTICE V

MASTERING **ALGEBRA 1**

AMERICAN MATH
═══ ACADEMY ═══

By H. TONG, M.Ed.

Math Instructor & Olympiad Coach

www.americanmathacademy.com

AMERICAN MATH ACADEMY

ESSENTIAL PRACTICE WORKBOOK FOR
MASTERING **ALGEBRA 1**

Writer: H.Tong
Copyright © 2021 The American Math Academy LLC.

Printed in United States of America.

ISBN: 9798514043422

Although the writer has made every effort to ensure the accuracy and completeness of information contained this book, the writer assumes no responsibility for errors, inaccuracies, omissions or any inconsistency herein. Any slighting of people, places, or organizations is unintentional.

Questions, suggestions, or comments, please email: americanmathacademy@gmail.com

TABLE OF CONTENTS

About the Author

Mr. Tong teaches at various private and public schools in both New York and New Jersey. In conjunction with his teaching, Mr. Tong developed his own private tutoring company. His company developed a unique way of ensuring his students' success on the math section of the SAT. His students, over the years, have been able to apply the knowledge and skills they learned during their tutoring sessions in college and beyond. Mr. Tong's academic accolades make him the best candidate to teach SAT Math. He received his master's degree in Math Education. He has won several national and state championships in various math competitions and has taken his team to victory in the Olympiads. He has trained students for Math Counts, American Math Competition (AMC), Harvard MIT Math Tournament, Princeton Math Contest, the National Math League, and many other events. His teaching style ensures his students' success. He personally invests energy and time into his students and sees what they're struggling with. His dedication to his students is evident through his students' achievements.

Acknowledgements

I would like to take the time to acknowledge the help and support of my beloved wife, my colleagues, and my students their feedback on this book was invaluable. Without their help, this book would not be the same. I dedicate this book to my precious daughters Vera and Nora who was, my inspiration to take on this project.

1. Which of the following numbers is irrational?

A) 3

B) −5

C) $\sqrt{7}$

D) $\sqrt{25}$

4. Which of the following equations matches the following sentence: Subtract 36 from 48 and then in divide by 3

A) $(48 - 36) - 3$

B) $(48 - 36) + 3$

C) $(48 - 36) \div 3$

D) $(36 - 48) \div 3$

2. Which of the following numbers is rational?

A) $\sqrt{15}$

B) π

C) 1.2345...

D) $\dfrac{3}{5}$

5. Which of the following is equal to $4\times5 - (6 + 3) - 63 \div 7$?

A) 0

B) 2

C) −4

D) −6

3. Which of the following is the reciprocal of $\dfrac{a}{b}$?

A) $\dfrac{a}{b}$

B) $\dfrac{b}{a}$

C) $-\dfrac{a}{b}$

D) $-\dfrac{b}{a}$

6. Which of following algebraic equations correctly represents this sentence: Thirty four is four times a number, increased by nine.

A) $43 = 4x - 9$

B) $34 = 4x + 9$

C) $4 = 9x + 34$

D) $9 = 34x + 4$

7. Which expression is equivalent to
$(5x^4 + 12x^2) - (12x^3 - 9x)$?

A) $5x^4 + 12x^2 - 12x^3 - 9x$

B) $5x^4 - 12x^3 + 12x^2 + 9x$

C) $5x^4 + 12x^2 - 12x^3$

D) $5x^4 - 12x^3 - 12x^2 - 9x$

8. If x is the greatest prime factor of 21 and y is the greatest prime factor of 57, what is the value of x + y?

A) 12

B) 15

C) 26

D) 30

E) 36

9. A school hired 30 teachers. This is 40% of the number of teachers it expects to have at the end of the next academic year. How many teachers does it expect to have next academic year?

A) 25

B) 35

C) 45

D) 75

10. Simplify $\dfrac{x^2 - 8x + 15}{x^2 - 9} \div \dfrac{x^2 - 4x - 5}{x^2 + 3x}$

A) $x - 1$

B) $\dfrac{x}{x+1}$

C) $x + 1$

D) $\dfrac{x+1}{x-1}$

11. In college, next year's tuition will increase by 15% per credit. If this year's tuition is $660 per credit, what will tuition be next year?

A) $650

B) $688

C) $745

D) $759

12. If $\begin{aligned} a &= 3^n \\ b &= 2^n \end{aligned}$, then find 18^n in terms of a and b.

A) $a \cdot b$

B) $a^2 \cdot b$

C) $a \cdot b^2$

D) $a^2 \cdot b^2$

American Math Academy

13.

$$(81x^8)^{\frac{1}{4}}$$

Which of the following equations is equivalent to the expression above?

A) x^2

B) $3x^2$

C) $\frac{1}{2}x^2$

D) $5x^2$

14. Simplify $\dfrac{\sqrt{a}}{2-\sqrt{a}}$

A) $\dfrac{2\sqrt{a}+a}{4-a}$

B) $\dfrac{\sqrt{a}-a}{4-a}$

C) $\dfrac{\sqrt{a}-a}{a}$

D) $\dfrac{2}{4-a}$

15. In a school, the ratio of the number of student in science class to the number of students in math class is 4:9. If the number of students in math class is 27, then what is the number of students in science class?

A) 6

B) 9

C) 12

D) 15

16. Two numbers have a ratio of 5 to 7. The larger number is 20 more than $\frac{3}{5}$ of the smaller number. Find the larger number.

A) 15

B) 20

C) 25

D) 35

17. Simplify $\left(\dfrac{a}{b}\right)\cdot\left(\dfrac{ab}{cd}\right)\cdot\left(\dfrac{cd}{a^2}\right)$?

A) a^2

B) $a^2 \cdot b$

C) $a \cdot b \cdot c \cdot d$

D) 1

18. If x is a positive number and $x^2 + x - 12 = 0$, what is the value of x?

A) 2

B) 3

C) 4

D) 6

19. x and y are integers.

$$-8 < x < 7$$
$$1 < y < 11$$

What is the maximum value of $x^2 + y^2$?

A) 109

B) 129

C) 149

D) 169

21. Melissa's final grade in a math course is 60% of her current grade, plus 30% of her final exam score. If her current grade is 80 and her goal is to get a final grade of 90 or higher score. Which of following is inequalities correctly represents for this situation?

A) $0.6 \cdot 90 + 0.3 \cdot x \geq 80$

B) $0.6 \cdot 30 + 0.8 \cdot x \geq 90$

C) $0.6 \cdot 80 + 0.3 \cdot x \leq 90$

D) $0.6 \cdot 80 + 0.3 \cdot x \geq 90$

American Math Academy

20. A fitness center has two membership plans. One charges a $23 membership fee and $7 per visit, and the other charges only $13 per visit. Which of the following systems of equations can be used to represent the fitness center membership plans?

A) $y = 7 + 23x$
   $y = 13$

B) $y = 23 + 7x$
   $y = 13x$

C) $y = 13$
   $y = 23 + 7x$

D) $y = 13x$
   $y = 23x + 7$

22. If $|2x - 7| > 9$, which of following is the graph of the equation?

A)
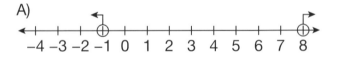

B)
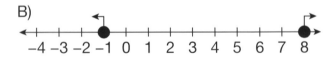

C)

D)

23. The population of a city increased from 270 thousand in 2017 to 360 thousand in 2018. Find the percent of increase.

A) $13.\overline{3}\%$

B) $23.\overline{3}\%$

C) $33.\overline{3}\%$

D) $43.\overline{3}\%$

24. Which statement is modeled by $3x + 7 > 10$?

A) The sum of 7 and 3 times x is at most 10.

B) Seven added to the product of 3 and x is greater than 10.

C) Three times x plus 7 is at least 10.

D) The product of 3 and x added to 7 is 10.

25. $\dfrac{2x-8}{3} - \dfrac{x+4}{4} = \dfrac{3}{2}$, find the value of x.

A) 12.2

B) 12.3

C) 12.4

D) 15.4

26. Simplify following equation.

For $x \neq -5$ and $x \neq -3$, which of the following is equivalent to

$$\dfrac{1}{\dfrac{1}{x+3} + \dfrac{1}{x+5}} = ?$$

A) $\dfrac{2x+8}{x^2+8x+15}$

B) $\dfrac{x^2+8x+15}{2x+8}$

C) $\dfrac{2x+8}{x^2-8x+15}$

D) $x^2 + 8x + 15$

27.

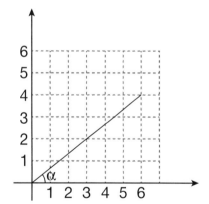

What is the equation of the function?

A) $y = 2x$

B) $y = \dfrac{2}{3}x$

C) $y = x + 7$

D) $y = 2x + 3$

American Math Academy

28. Simplify $\dfrac{1}{x-1}+\dfrac{1}{x+1}$

   A) $x-1$

   B) $x+1$

   C) 1

   D) $\dfrac{2x}{x^2-1}$

29. Simplify $\dfrac{x^2-7x+12}{x^2-9x+20}$

   A) $\dfrac{x-3}{x-4}$

   B) $\dfrac{x-3}{x-5}$

   C) $\dfrac{x-4}{x-3}$

   D) $\dfrac{x-5}{x-3}$

30. $\dfrac{1}{a}=\dfrac{1}{b}+\dfrac{1}{c}$ Find b in terms of a and c.

   A) $b=\dfrac{ac}{c-a}$

   B) $b=\dfrac{ac}{a-c}$

   C) $b=ac$

   D) $b=\dfrac{1}{c-a}$

31. Simplify $\dfrac{18}{6}-2\div4$

   A) 2

   B) 4

   C) $\dfrac{5}{2}$

   D) $\dfrac{1}{2}$

32. Simplify $\sqrt{\dfrac{34-2}{2}}+4^3-32$

   A) 16

   B) 28

   C) 32

   D) 36

33. Simplify $2a + 3b + a - 2b$

   A) $2a + b$

   B) $3a + b$

   C) $3a$

   D) $a + 2b$

American Math Academy

34. Simplify $2^2a^3 + 4b^3 + a^3 - 2b^3$

A) $5a^3 + 2b^3$

B) $5a^3 - b^3$

C) $a^3 + b^3$

D) $a^3 - b^3$

37. Solve $\sqrt{21} + \sqrt{84}$

A) $2\sqrt{11}$

B) $2\sqrt{21}$

C) $3\sqrt{21}$

D) $4\sqrt{11}$

35. Simplify $\sqrt{a^6} \cdot \sqrt{a^8}$

A) a^5

B) a^6

C) a^7

D) a^8

38. Simplify $(3x^6)^5$

A) $243x^{30}$

B) $9x^{11}$

C) $3x^{30}$

D) $81x^{30}$

36. Solve $\sqrt{75}$

A) $3\sqrt{5}$

B) $5\sqrt{3}$

C) $4\sqrt{5}$

D) $3\sqrt{3}$

39. Simplify $\dfrac{(x^2y^4)^4}{x^7y^8}$

A) x^1y^3

B) x^8y^8

C) x^8y^1

D) x^1y^8

American Math Academy

40. Find the prime factorization for 18.

A) $2^2 \cdot 3^2$

B) $2 \cdot 3^2$

C) $2 \cdot 3$

D) $2^2 \cdot 3$

41. Simplify $\dfrac{4x - 2y}{2x - y}$

A) 1

B) 2

C) $2x - y$

D) $x - y$

42. Simplify $\dfrac{\frac{3x^2}{4y^3}}{\frac{x^{-1}}{y^8}}$

A) $\dfrac{x^3 \cdot y^5}{4}$

B) $\dfrac{x^3 \cdot y^5}{5}$

C) $\dfrac{3x^3 \cdot y^5}{4}$

D) $\dfrac{3x^5 \cdot y^3}{4}$

43. Write the answer in scientific notation form
0.00056712

A) 5.6712×10^{-1}

B) 5.6712×10^{-2}

C) 5.6712×10^{-3}

D) 5.6712×10^{-4}

44. Write the answer in scientific notation form
875

A) 8.75×10^2

B) 8.75×10^1

C) 8.75×10^0

D) 8.75

45. Write the answer in scientific notation form
$(4^3 \times 10^{-4})(2 \times 10^{-8})$

A) 1.28×10^{-5}

B) 1.28×10^{-10}

C) 1.28×10^{-8}

D) 1.28×10^{10}

American Math Academy

46. Find the ratio of the shaded portion to the unshaded portion in the following figure.

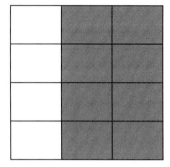

A) $\frac{2}{3}$

B) $\frac{3}{2}$

C) $\frac{4}{5}$

D) 2

47. If $x^{\frac{1}{4}} = 2$ and $x \cdot y = 64$, then what is the value of y?

A) 1

B) 2

C) 3

D) 4

48. What is the least common multiple of 7 and 8?

A) 14

B) 21

C) 28

D) 56

49. Simplify $\frac{\sqrt{x}}{2+\sqrt{x}}$?

A) $\frac{2\sqrt{x}-x}{4-x}$

B) $\frac{\sqrt{x}-x}{4-x}$

C) $\frac{\sqrt{x}-x}{x}$

D) $\frac{2}{4-x}$

50. The average of five consecutive positive integers is 28. What is the greatest possible value of one of these integers?

A) 24

B) 26

C) 30

D) 32

American Math Academy

Parentheses

Do any operations in parentheses

$$36 + 4(8 - 3) - 2^3 \div 4$$

$$36 + 4 \times 5 - 2^3 \div 4$$

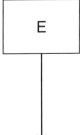

Exponents

Do any Exponents

$$36 + 4 \times 5 - 2^3 \div 4$$

$$36 + 4 \times 5 - 8 \div 4$$

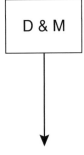

Divide and Multiply rank equally (Always work, solve, or move etc. left to right)

$$36 + 4 \times 5 - 8 \div 4$$

$$36 + 20 - 2$$

Add and subtract rank equally (Always work, solve, or move etc. left to right)

$$36 + 20 - 2$$

$$56 - 2$$

$$54$$

ORDER OF OPERATIONS EXERCISES

Solve each of following. Be sure to follow **PEMDAS**.

1) $6(3 - 3) + 5^2$

2) $3 (2 \times 3 - 6) + 3^3$

3) $52 - 12 \times 4 + 11$

4) $6(7 \times 3 - 6) - 2^3$

5) $6(22 - 7 \times 3) + 7 \times 3 - 2^3$

6) $25 + (-3) \times 4^2 \div 12$

7) $\{125 \div 5^3 \times 3 - 15\} + 25°$

8) $\{32 \div 2^3 \times 3 + 15\} - 12$

9) $\dfrac{3^4(2^4)}{3 \times 16}$

10) $\dfrac{8^2 - (-12 + 16)}{3(4^2 - 8) - 4}$

11) $\dfrac{32 - (8 - 8)}{2^5}$

12) $\dfrac{5(10 - 6) + 3^2(14 - 4)}{2^2 + 2}$

13) $12 + 16 - 3(\sqrt{49} + 3)$

14) $\dfrac{25 - 10}{5} - \sqrt{36} + 15$

15) $\{\sqrt{64} \div 2^3 \times 5 + 20\} - 10$

16) $100 - 5^2 + \dfrac{\sqrt{36}}{\sqrt{9}}$

FRACTIONS AND OPERATIONS
WITH FRACTIONS

Key Notes:

Fractions: Fractions are numbers that can be in the form $\frac{A}{B}$ and B is not equal to zero.

Adding Fractions: When you add fractions, if they have same denominator, then you add the numerators while keeping the denominator the same.

If the fractions have different denominators:
- Find the smallest multiple (LCD) of both numbers.
- Rewrite the fractions as equivalent fractions with the LCD as the denominator.

Subtracting Fractions: When you subtract fractions that have the same denominators, you subtract only the numerators and keep the denominator the same.

If the fractions have different denominators:
- Find the smallest multiple (LCD) of both numbers.
- Rewrite the fractions as equivalent fractions with the LCD as the denominator.

Multiplying Fractions:
- Multiply the numerators
- Multiply the denominators
- Simplify the fraction if needed

Dividing Fractions:
- Flip the divisor
- Multiply the first fraction by that reciprocal.
- Simplify the fraction if need.

Find the value of each expression in lowest terms.

1) $-\dfrac{1}{3} + \dfrac{1}{5}$

2) $\dfrac{1}{4} - \dfrac{1}{12}$

3) $\dfrac{3}{5} - \dfrac{4}{7}$

4) $\dfrac{7}{49} + \dfrac{6}{36}$

5) $1\dfrac{1}{4} + 2\dfrac{1}{3}$

6) $3\dfrac{1}{2} - 4\dfrac{1}{5}$

7) $\dfrac{1}{8} \times \dfrac{16}{10}$

8) $3\dfrac{1}{8} \times \dfrac{1}{9}$

9) $\dfrac{1}{18} \div \dfrac{9}{4}$

10) $5\dfrac{1}{2} \div \dfrac{22}{3}$

11) $\dfrac{1}{18} \times 72$

12) $1\dfrac{1}{17} \div \dfrac{1}{34}$

13) $-\dfrac{1}{12} \div -2\dfrac{1}{4}$

14) $33 \div \dfrac{22}{3}$

15) $\dfrac{1}{12} + 8$

16) $\dfrac{\dfrac{1}{2}}{\dfrac{3}{5}}$

American Math Academy

INTEGERS AND OPERATIONS
WITH INTEGERS

Key Notes:

Whole Numbers: A number without a decimal or fraction.

- The whole numbers are the numbers 0, 1, 2, 3, 4, 5, 6, 7 and so on
- 0 is the smallest whole number
- Negative numbers are not considered whole numbers.

Natural Numbers: The set of counting numbers, starting at 1 and going up.

- The natural numbers are the numbers 1, 2, 3, 4, 5, 6, 7 and so on.
- 1 is the smallest natural number.
- Negative numbers are not considered natural numbers.

Integers: The set of all whole numbers from positive to negative infinity.

Positive Integers: Whole numbers greater than zero.

Negative Integers: Whole numbers less than zero.

Helpful Hints:

$(+) \cdot (+) = +$	$(+) / (+) = +$
$(+) \cdot (-) = -$	$(+) / (-) = -$
$(-) \cdot (+) = -$	$(-) / (+) = -$
$(-) \cdot (-) = +$	$(-) / (-) = +$

Reciprocal: The reciprocal is the multiplicative inverse of a number.

Rational numbers: A rational number is a number that can be in the form $\frac{A}{B}$ and B is not equal to zero.

- Every whole is a rational number.
- Every natural number is a rational number.
- Every integer number is a rational number.
- Every repeating decimal number is a rational number.
- Square roots of all positive integers are rational numbers.

Irrational Numbers: Any real number that cannot be written in fraction form or has non-repeating decimals.

- If a square root is not a perfect square, then it is considered an irrational number.
- Irrational numbers have no exact decimal equivalents.

INTEGERS AND OPERATIONS WITH INTEGERS EXERCISES

Simplify each of the following.

1) $4 - (-8)$

2) $-12 - (20)$

3) $4^2 \div (-\sqrt{36})$

4) $-25 + (-34) - (-9)$

5) $-4^2 \div -(-24)$

6) $-(-15) + \left(\dfrac{1}{2}\right) - (14)$

7) $\dfrac{85 \div 17}{-15 + 10}$

8) $\dfrac{-8 \div 2^3}{-(-15 + 8)}$

9) $(-3)(-5)(-4)$

10) $\dfrac{-1}{2} \times \dfrac{-2}{3} \times \dfrac{-3}{4}$

Find the reciprocal for each of the following.

11) $\dfrac{a}{b}$

12) $-\dfrac{3}{5}$

13) $\dfrac{-\frac{1}{4}}{-\frac{1}{5}}$

14) $-1\dfrac{4}{5}$

Find the opposite for each of the following.

15) $-5(-8)$

16) $-12(-2)(-1)$

Key Notes:

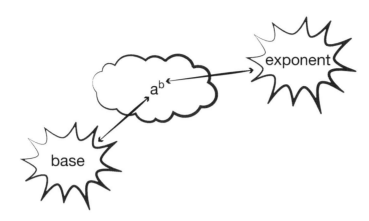

Product Rule: When multiplying exponents with the same base, always add the powers and keep the base the same.

$$a^x \times a^y = a^{x+y}$$

Quotient Rule: When dividing exponents with the same base, always subtract the powers and keep the base the same.

$$\frac{a^x}{a^y} = a^{x-y}$$

Power to power Rule: When you have one base but more than one power, keep the base and multiply all powers together.

$$a^{x^y} = a^{x \cdot y}$$

Inverse: When you have a negative power always take the reciprocal of the base and make the power positive.

$$a^{-x} = \frac{1}{a^x}$$

Zero Exponents: Any number to the power of zero is always 1.

$$a^0 = 1$$

Simplify each of the following.

1) $2^3 \times 2^4 \times 2^8$

2) $3^{-4} \times 3^{-12} \times 3^{16}$

3) $\dfrac{6^{-3}}{6^{-4}}$

4) $\dfrac{25 \times 125}{5^6}$

5) $\left(\dfrac{x^3}{y^6}\right)^0$

6) $(2xy^3)^3$

7) $(3xy^{-5})^{-2}$

8) $\left(\dfrac{x^3}{3}\right)^{-2}$

9) $\dfrac{3^4 \times 3^{-5} \times 3^{12}}{3^{-7}}$

10) $\dfrac{36x^2y^3}{12x^4y^2}$

11) $\dfrac{3x^{-3}y^{-4}}{9x^{-5}y^{-6}}$

12) $\dfrac{\dfrac{xy^3}{1}}{\dfrac{1}{x^3y^{-4}}}$

13) $\dfrac{y^3}{y^{-8}} \times y^{-5}$

14) $2^x \times 2^x \times 2^x$

15) $3^x + 3^x + 3^x$

16) $\dfrac{4^x}{16^{3x}} \times 32^x$

Key Notes:

Absolute Value: The distance of integers from zero on the number line.

- If |x| = a, then either x = a or x = –a
- If |ax + b| = k, then either ax + b = k or ax + b = –k
- If |ax + b| < k, then either ax + b < k or ax + b > –k
- |ax + b| ≤ k, then either ax + b ≤ k or ax + b ≥ –k

Inequality Symbols

Symbol	Meaning	Number Line	Circle
<	Less than	←——o	Open circle
>	Greater than	o——→	Open circle
≤	Less than or equal to (At least)	←——●	Close circle
≥	Greater than or equal to (At most)	●——→	Close circle

Solve and simplify each of the following.

1) $|3x - 6| = 9$

2) $|x - 5| = -10$

3) $4|2^3 - 7|$

4) $\frac{1}{11} |3^3 - 4^2|$

5) $\frac{|2^3 + 2^2|}{2^4}$

6) $3x + 5 > 9$

7) $|2x - 10| < 12$

8) $|3x - 6| > 18$

9) $1 - \frac{2x}{3} < x - 4$

10) $3x - 1 \geq 2(x - 5)$

11) $\frac{2x + 6}{5} < -4$

12) $\frac{3}{4x} < \frac{5}{x + 6}$

13) $\frac{2x - 7}{4} \leq \frac{3x + 5}{6}$

14) $4|x - 8| - 6 > 18$

15) $|3x - 5| = 5x$

16) $\frac{1}{3}|x| - 3 = 10$

American Math Academy

LAWS OF RADICALS

Key Notes:

The symbol for square root is $\sqrt{}$

$\sqrt[x]{a^x} = a$, When $a \neq 0$ and $x > 1$.

$\sqrt[x]{a^x} = a$, When $a < 0$ and x is an even number.

$\sqrt[x]{ab} = \sqrt[x]{a} \cdot \sqrt[x]{b}$

$\sqrt[x]{\dfrac{a}{b}} = \dfrac{\sqrt[x]{a}}{\sqrt[x]{b}}$

$\sqrt[x]{\sqrt[y]{a}} = \sqrt[x \cdot y]{a}$

Simplify each of the following.

1) $\sqrt{144}$

2) $\sqrt{16} + \sqrt{25} - \sqrt{36}$

3) $\sqrt{800} + \sqrt{200}$

4) $\sqrt[3]{\dfrac{8}{27}}$

5) $\dfrac{\sqrt{25}}{\sqrt[3]{27}}$

6) $\sqrt{18} - \sqrt{72}$

7) $\sqrt[3]{8,000,000}$

8) $\dfrac{6\sqrt{5}}{\sqrt{3}}$

9) $\sqrt{a} \cdot \sqrt{a} \cdot \sqrt{a}$

10) $\dfrac{1}{\sqrt{x}} \cdot \dfrac{1}{\sqrt{x}} \cdot \dfrac{1}{\sqrt{x}} \cdot \dfrac{1}{\sqrt{x}}$

11) $\sqrt{121} + \sqrt{64} - \sqrt{144}$

12) $3\sqrt{3} - 2\sqrt{3} + \sqrt{3}$

13) $2\sqrt{a} \cdot \dfrac{1}{\sqrt{a}} \cdot \sqrt{a}$

14) $2\sqrt{20} - \sqrt{5}$

15) $3\sqrt{45} - 9\sqrt{5} + \sqrt{75}$

16) $3\sqrt{3} + \sqrt{27} + \sqrt{12}$

American Math Academy

COORDINATE PLANE

Key Notes:

Coordinate Plane: A coordinate system formed by the intersection of a vertical number line, called the y–axis, and a horizontal number line, called the x–axis.

Origin: A beginning or starting point and lines that intersect each other at zero (0,0).

Quadrant: One of four regions into which the coordinate plane is divided by x– and y–axis. These regions are called the quadrants.

Ordered pair: A pair of numbers that can be used to locate a point on a coordinate plane. The order of the numbers in a pair is important, and the x–axis always comes before the y–axis.

X – coordinate: The first number in an ordered pair is called the x–coordinate, and it is the first value in an ordered pair.

Y – coordinate: The second number in an ordered pair is called the y–coordinate. It is the second value in an ordered pair.

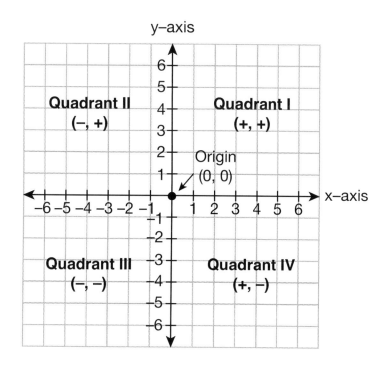

COORDINATE PLANE EXERCISES

Name the coordinates for each given point. Give the quadrant for each point.

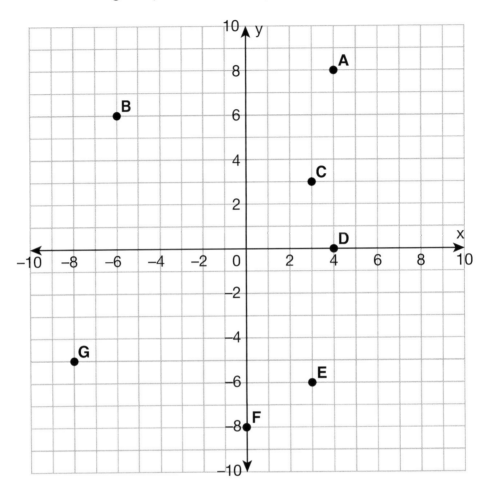

Coordinates	Point	Quadrant
	A	
	B	
	C	
	D	
	E	
	F	
	G	

AMERICAN MATH
ACADEMY

Key Notes:

Factors: A number that will divide into another number without a remainder.

Prime Factorization: The form of a number written as the product of its prime factors.

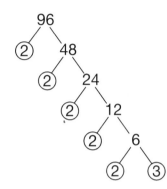

The prime factorization of 96 is

$2 \cdot 2 \cdot 2 \cdot 2 \cdot 2 \cdot 3 = 2^5 \cdot 3$

Prime Numbers: A number that has only two factors, 1 and itself.

Composite Number: A number that has more than two factors.

Least Common Multiple (LCM): The least common multiple of two numbers is the smallest integer that is a multiple of both numbers.

Example: Find the LCM of 3 and 5.

Solution:
Multiples of 3: 0, 3, 6, 9, 12, 15...
Multiples of 5: 0, 5, 10, 15...
15 is the least common multiple of 3 and 5.

Greatest Common Factor (GCF): The largest number that is a factor of two or more numbers.

Example: Find the GCF of 12 and 18.

Solution:
Factors of 12: 1, 2, 3, 4, 6, 12.
Factors of 18: 1, 2, 3, 6, 9, 18.
6 is the greatest common factor of 12 and 18.

FACTORS & MULTIPLES
(GCF AND LCM) EXERCISES

Circle the numbers that are prime

1	2	3	4	5
6	7	8	9	10
11	12	13	14	15
16	17	18	19	20
21	22	23	24	25

Circle the numbers that are composite

1	2	3	4	5
6	7	8	9	10
11	12	13	14	15
16	17	18	19	20
21	22	23	24	25

Find the greatest common factor (GCF) and least common multiple (LCM) of each pair of **numbers**

	GCF	LCM
15, 45		
16, 24		
18, 48		
5, 45		
11, 88		
5, 10, 15		
16, 24, 36		

Key Notes:

$$A \times 10^n \longleftarrow \text{integer}$$

Always use base 10

$$1 \leq |A| < 10$$

Where A is a number greater than or equal to 1 but less than 10

Examples:

1. $0.00015 = 1.5 \times 10^{-4}$

2. $156,000 = 1.56 \times 10^5$

3. $685 \times 10^{-3} = 6.85 \times 10^{-1}$

4. $123,000,000 = 1.23 \times 10^8$

5. $12 = 1.2 \times 10^1$

6. $5 = 5 \times 10^0$

SCIENTIFIC NOTATION EXERCISES

Write each exponent in scientific notation form. Show your work.

1. $125 \times 10^5 =$ _____

2. $0.0123 \times 10^5 =$ _____

3. $0.98 \times 10^{-6} =$ _____

4. $145{,}000 \times 10^3 =$ _____

5. $0.0457 \times 10^{-9} =$ _____

6. $86.9 \times 10^{-12} =$ _____

Write each exponent in standard form. Show your work.

7. $967 \times 10^4 =$ _____

8. $0.0457 \times 10^{-1} =$ _____

9. $21.9 \times 10^3 =$ _____

10. $1.5 \times 10^0 =$ _____

11. $0.04 \times 10^6 =$ _____

12. $3.45 \times 10^7 =$ _____

Write each number in scientific notation form. Show your work.

13. $1.8 =$ _____

14. $0.0444 =$ _____

15. $3.784 =$ _____

16. $89{,}000{,}000 =$ _____

17. $3.67 =$ _____

18. $234{,}000 =$ _____

RATIO, PROPORTIONS AND VARIATIONS

Key Notes:

Ratio: A ratio is a comparison or relation between two quantities.

Note: The ratio of a and b is written as a to b, or $\frac{a}{b}$.

Proportions: A proportion is an equation that shows two equivalent ratios.

if $\frac{a}{b} = \frac{c}{d}$, then, by cross multiplication, ad = bc.

Direct variation: A direct variation is a relationship between two quantities of x and y that can be written as the following form:

- y = kx, where k is a constant of veriation and k does not equal zero.

Inverse variation: an inverse variation is a relationship between two quantities of x and y that can be written as the following form:

- xy = k, where k is a constant of variation and K does not equal zero.

RATIO, PROPORTIONS AND VARIATIONS EXERCISES

Write each of the following ratios as fractions in simplest form

1) 20 books to 30 pencils

2) 8 bananas to 35 apples

3) 17 to 68

4) 72 : 66

Determine whether each of the following is an equivalent ratio.

5) $\dfrac{1}{2}$ and $\dfrac{6}{4}$

6) $\dfrac{1}{2}$ and $\dfrac{12}{24}$

7) $\dfrac{2}{18}$ and $\dfrac{6}{54}$

8) $\dfrac{7}{35}$ and $\dfrac{90}{120}$

Find the missing number in each of following.

9) $\dfrac{x}{6} = \dfrac{6}{54}$

10) $\dfrac{3}{x} = \dfrac{7}{54}$

11) $\dfrac{-x}{5} = \dfrac{15}{25}$

12) $\dfrac{x+5}{8} = \dfrac{9}{4}$

13) $\dfrac{x}{12} = \dfrac{\frac{1}{4}}{\frac{3}{5}}$

14) $x = \dfrac{\frac{1}{8}}{\frac{16}{12}}$

15) The ratio of girls to boys in math class is 3 : 5. If there are 32 total students in class, how many boys are there?

16) Find y when x = 3, if y varies directly as x and y = 20 when x = 5.

17) If y varies inversely as x and x = 12 when y = 60, find y when x = 18.

AMERICAN MATH
ACADEMY

Key Notes:

Percent: A fraction that is a partition of 100.

$$\text{Perecent change} = \frac{\text{amount of change}}{\text{original amount}}$$

Example: $15\% = \dfrac{15}{100}$

Types of Percent Questions

Examples:

1) What is 20% of 40?

Solution: $P = \dfrac{20 \times 40}{100} = 8$

2) What percent of 20 is 80?

Solution: $\dfrac{20n}{100} = 80$, then $n = \dfrac{80 \times 100}{20}$, $n = 400$

3) 20 is 25% of what number?

Solution: $20 = \dfrac{25m}{100}$, then $m = \dfrac{20 \times 100}{25}$

$$m = 80$$

Use the data in each table below to find the unit rate.

1.

Days	1	2
Miles	30	60

Unite Rate: _____ miles/day

2.

Hours	5	10
$	20	40

Unite Rate: _____ $/hours

3.

Book	7	21
Pages	35	105

Unite Rate: _____ pages/book

Solve each problem for unknown

4. 10 is as 25 % of what number?

5. What is 45% of 40?

6. What % of 30 is 6?

7. What is 10% of 90?

8. A cleaning company charges $480 to clean 24 classrooms.

What is the company's price for cleaning a single class?

9. The price of a book has been discounted 20%. The sale price is $45. What is the original price?

10. A microwave originally priced at $40 is decreased in price by 25%. What is the sale price?

1. If $x = \dfrac{4^3}{2^7}$, then find x

A) 2

B) 4

C) $\dfrac{1}{2}$

D) $\dfrac{1}{4}$

2. Simplify $\sqrt[3]{64}$

A) 2

B) 4

C) 8

D) 16

3. Simplify $\sqrt{25} - \sqrt{36} + \sqrt{100}$.

A) 2

B) 3

C) 6

D) 9

4. Simplify $\sqrt{(16)^2} + \sqrt{2^2} - (-2)^3$

A) 15

B) 18

C) 22

D) 26

5. If $x\sqrt{25} = 45$, then find x

A) – 9

B) 9

C) 12

D) – 12

6. Simplify $\dfrac{\sqrt{250}}{\sqrt{10}} \cdot \sqrt{49}$

A) 10

B) 15

C) 35

D) 45

7. If $x = 1 + \sqrt{2}$ and $y = 1 - \sqrt{2}$, then $x \cdot y = ?$

A) $-2\sqrt{3}$

B) –1

C) $2\sqrt{3}$

D) 1

8. Which of following is equal to $\dfrac{8^2 \cdot 16^3}{2^{10}}$?

A) 2^4

B) 2^6

C) 2^8

D) 2^{10}

American Math Academy

9. For all values of x, which of the following is equal to $x^{\frac{2}{7}}$?

A) x^2

B) $\sqrt[6]{x^7}$

C) $\sqrt[7]{x^2}$

D) x^7

10. If $x^5 = 32$, then find x^2.

A) 1

B) 2

C) 4

D) 8

11. Last night, Melisa spent $1\frac{3}{8}$ hours studying her math homework. John studied his math homework for $\frac{1}{4}$ as many hours as Melisa did. How many hours did John spend on his homework?

A) $\frac{9}{32}$

B) $\frac{1}{2}$

C) $\frac{3}{8}$

D) $\frac{11}{32}$

12. Jennifer has $258 in her bank checking account. How much does she have in her checking account after she makes a deposit of $140 and a withdrawal of $128?

A) $270

B) $372

C) $385

D) $512

13. If the sum of three consecutive integers is 96, then what is the smallest integer?

A) 18

B) 21

C) 29

D) 31

14. If $4^{x-4} = 8^{x-6}$, then what is the value of x?

A) 5

B) 10

C) −5

D) −10

American Math Academy

15. What is the solution to the equation below?

$$\frac{2x-7}{5} = \frac{2x-9}{3}$$

A) 2

B) 3

C) 5

D) 6

18. If $x > 0$ and $(3x - 5)^2 = 49$, then what is the value of x?

A) 3

B) 4

C) 5

D) 6

16. If $x = \dfrac{2^3}{\sqrt{64}}$, then x = ?

A) 1

B) $\dfrac{1}{2}$

C) 2

D) 4

19. If $x + 2y = 3$, then find $2^x \cdot 4^y$

A) 2

B) 4

C) 6

D) 8

17. If $y - 2x = 10$, then which of following is equal to 4x?

A) $y - 5$

B) $2y - 5$

C) $2y - 20$

D) $2y + 20$

20. Melisa has 24 pens and wants to give ¾ of them to a friend and keep the rest for herself. How many pens would her friend get?

A) 9

B) 12

C) 18

D) 21

American Math Academy

21. If $20 = 5\left(\dfrac{x}{3} - 4\right)$, then what is the value of x?

A) 24

B) $\dfrac{1}{24}$

C) 48

D) $\dfrac{1}{48}$

23. The formula $F = \dfrac{9}{5}C + 32$ shows the temperature in degrees Fahrenheit for a given temperature in degrees Celsius C. Find F when C = 60.

A) 40F

B) 80F

C) 96F

D) 140F

22. If a and b are positive integers and $\sqrt{a} = b^2 = 4$, then which of following is the value of a − b?

A) 6

B) 9

C) 12

D) 14

1. How many solutions does the system of equations shown below have?

$$a = 3b + 6$$
$$3a - 9b = 18$$

A) Zero

B) 1

C) 2

D) Many/Infinity

2. If $x > 0$, what is the value of x in $|2x - 5| = 7$?

A) –3

B) 3

C) 6

D) –6

3. Which of following is equal to $81x^2 - 4y^2$?

A) $(9x - 2y)(9x + 2y)$

B) $9x + 4y$

C) $9x - 4y$

D) $(9x - 2y)(9x - 2y)$

4. If $5a - 3b = -6$ and $b = 6$, then find a.

5. If $x^3 = 27$, then find x^2.

A) 3

B) 9

C) 27

D) 81

6. If the ratio of $\frac{1}{3} : \frac{1}{b}$ is equal to $\frac{1}{4} : \frac{1}{2}$, what is the value of b?

A) 2

B) $\frac{3}{2}$

C) $\frac{2}{3}$

D) 3

7. If $3^{2x-4} = 27^{x-6}$, then what is the value of x?

A) 10

B) 12

C) 14

D) −14

8. If $\frac{3x-1}{2x+1} = \frac{5}{3}$, then what is the value of x?

A) 6

B) −6

C) 8

D) −8

9. If $x = \frac{3^3}{\sqrt{81}}$, then find x.

A) 1

B) $\frac{1}{3}$

C) 3

D) 6

10. A line is represented by the equation $x + ay - 6 = 0$. If the slope of the equation is $\frac{1}{3}$, what is the value of a?

A) 3

B) $-\frac{1}{3}$

C) −3

D) $\frac{1}{6}$

11. If line L passes through the coordinate points (−1, 5) and (2, 8), what is the slope of line L?

A) 1

B) −1

C) 2

D) −2

American Math Academy

12.
$$x - 2y = 12$$
$$3x - 4y = 18$$

In the system of equation above, what is the valve of x?

A) 1

B) 6

C) –6

D) 8

13. If $\dfrac{2x-1}{3} - \dfrac{x}{2} = \dfrac{3}{2}$, What is the value of x?

A) 9

B) 11

C) 14

D) 15

14. Find the ratio of $\dfrac{1}{3}$ to $\dfrac{4}{3}$

A) $\dfrac{1}{6}$

B) $\dfrac{1}{3}$

C) $\dfrac{5}{7}$

D) $\dfrac{1}{4}$

15. Evaluate $4^{15} \times 4^2 = 4^{-n} \times 4^m \times 4^7$, then find $m - n$

A) 10

B) 12

C) 14

D) 16

16. Which of following is equivalent to $x^{\frac{3}{4}}$?

A) $\sqrt{x}$

B) $\sqrt[3]{x^4}$

C) $\sqrt[4]{x^3}$

D) x

17. Which of following has the same value as $\sqrt{3} \cdot \sqrt{16}$?

A) $4\sqrt{3}$

B) $4\sqrt{2}$

C) $3\sqrt{2}$

D) 12

18. Simplify $\sqrt[3]{0.027}$

A) 1

B) 0.3

C) –0.3

D) 3

American Math Academy

SOLVING 2 - STEP EQUATIONS

Key Notes:

- Isolate the variable on one side of equation
- Multiply or divide to solve for the variable
- Check your answer

Example: $2x + 10 = 20$

Solution: $2x + 10 = 20$

$$\underline{+\qquad -10 = -10}$$

$$2x = 10$$

$$x = 5$$

$$2(5) + 10 = 10 + 10 = 20 \text{ (check)}$$

Variable: A letter, that represents an unknown number. Any letter can be used as a variable.

Coefficient: A number placed before a variable.

Constant: Any real number used in an expression or equation.

Terms: Terms are separated by positive or negative signs in an expression.

Like terms: Like terms are terms that have the same variables with the same exponents.

Coefficient   Variable

$$4x + 8 = 7$$

Operator   Constants

Solve each of the following equations.

1) $5x - 10 = 15$

2) $3x - 15 = 30$

3) $4x - 5 = -6$

4) $12x + 24 = 48$

5) $-6x + 10 = -20$

6) $7x - 28 = 0$

7) $\dfrac{x}{3} - 5 = 10$

8) $-\dfrac{3x}{8} - 2 = 4$

9) $\dfrac{3x}{7} - \dfrac{1}{14} = 9$

10) $\dfrac{x}{-7} - 7 = 9$

11) $3x - (-12) = -9$

12) $8x + 10 = 50$

13) $\dfrac{x}{3} - \dfrac{1}{4} = \dfrac{1}{12}$

14) $\dfrac{x}{7} - (-6) = -1$

15) $\dfrac{-1}{3} - \dfrac{x}{2} = 7$

16) $12\left(\dfrac{x}{3} - \dfrac{1}{4}\right) = 9$

17) $\dfrac{x}{2}\left(\dfrac{1}{6} - \dfrac{1}{3}\right) = 5$

18) $3 + \dfrac{x}{7} = -2$

American Math Academy

SOLVING EQUATIONS WITH VARIABLE IN BOTH SIDES

Key Notes:

- First use distributive property rule to remove any parentheses
- Add or subtract terms to get the variable on one side
- Multiply or divide to solve for the variable
- Check your answers

Example:

$2(3x + 10) = 3x + 12$

Solution:

$6x + 20 = 3x + 12$

$6x - 3x + 20 = 12$

$3x + 20 = 12$

$3x = 12 - 20$

$3x = -8$

$x = \dfrac{-8}{3}$

Check:

$6x + 20 = 3x + 12$

$6\left(\dfrac{-8}{3}\right) + 20 = 3\left(\dfrac{-8}{3}\right) + 12$

$-16 + 20 = -8 + 12$

$4 = 4$

SOLVING EQUATIONS WITH VARIABLE IN BOTH SIDES EXERCISES

Solve each of the following equations.

1) $2x - 10 = x + 5$

2) $5x - 15 = 4x + 12$

3) $4x - 5 = -6 + 3x$

4) $-3x + 24 = -5x - 12$

5) $-6x + 10 = -20 + 10x$

6) $x - 28 = 3x$

7) $\frac{x}{3} - 5 = \frac{x}{4} + 2$

8) $-\frac{1}{2} - 2x = 3x - \frac{1}{2}$

9) $\frac{3}{7} - \frac{x}{7} = \frac{9}{7} + \frac{2x}{7}$

10) $\frac{x}{-1} - 7 = -x + 5$

11) $3x - (-2x) = -12 - 3$

12) $2(x + 10) = 5(x - 4)$

13) $\frac{x}{10} - \frac{1}{5} = \frac{x}{5}$

14) $2x - (-6) = -1(x - 8)$

15) $\frac{-12}{2} - \frac{2x}{2} = 2x$

16) $12\left(\frac{x}{3} - \frac{1}{4}\right) = 2x - 3$

17) $\frac{x}{3}\left(\frac{1}{6} - \frac{1}{3}\right) = \frac{x}{2}$

18) $3 + \frac{x}{3} = -5 + x$

PROPERTIES OF ALGEBRAIC EQUATIONS AND SIMPLIFYING EQUATIONS

Key Notes:

- Commutative Property of Addition: a+b=b+a

  Example: 2 + 3 = 3 + 2

- Commutative Property of Multiplication a·b=b·a

  Example: 3 · 4 = 4 · 3

- Associative Property of Addition: a + (b + c)= (a + b) + c

  Example: 4 + (5 + 6) = (4 + 5) + 6

- Associative Property of Multiplication: a · (b · c) = a(· b) · c

  Example: 5 · (7 · 9) = (5 · 7) · 9

- Distributive Property: a(b + c) = ab + ac

  Example: 2(4 + 6) = 2 · 4 + 2 · 6

- Additive Identity: 0 + a = a + 0

  Example: 0 + 1 = 1 + 0

- Multiplicative Identity: 1 · a = a · 1

  Example: 1 · 3 = 3 · 1

- Additive Inverses: a + (−a) = 0

  Example: 8 + (−8) = 0

- Multiplicative Inverses: $a \cdot \dfrac{1}{a} = 1$

  Example: $5 \cdot \dfrac{1}{5} = 1$

- Zero Property of Multiplication: a · 0 = 0

  Example: 7 · 0 = 0

AMERICAN MATH
ACADEMY

Simplify each of the following.

1) $3(8 + 5)$

2) $5(7 + 2)$

3) $a(b + c)$

4) $4(x + 3)$

5) $-3(3x - 5)$

6) $-2(-x - 5)$

7) $-2(-x - 4)$

8) $4(x + 7)$

9) $\frac{1}{2}(x - 12)$

10) $\frac{x}{3}(x - 6)$

11) $3x(a + 2b)$

12) $\frac{1}{3}(6x - 9y)$

13) $-3(x - 2y + 4z)$

14) $(3x - 6) - (2x - 8)$

15) $(5x - 1) \cdot (2x + 1)$

16) $(a - b) \cdot (a + b)$

17) $(x + y) \cdot (x - y)$

18) $(a^2 - b^2) \cdot (a^2 + b^2)$

19) $\left(\frac{1}{x} + x\right) \cdot \left(\frac{1}{x} - x\right)$

20) $a^2 (a^2 - b^2)$

American Math Academy

SOLVING EQUATIONS INVOLVING PARALLEL AND PERPENDICULAR LINES EXERCISES

Key Notes:

Parallel lines (No solution):

The equations have the same slope and different y–intercepts. The graphs are parallel.

$$\left.\begin{array}{l} d_1 : y = m_1 x + n_1 \\ d_2 : y = m_2 x + n_2 \end{array}\right\} \quad d_1 \mathbin{\!/\mkern-5mu/\!} d_2 \Longleftrightarrow m_1 = m_2$$

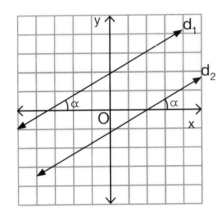

– Same slope $(m_1 = m_2)$
– Different y–intercepts

Example:

2x + y = 3
6x + 3y = 12

Perpendicular lines:

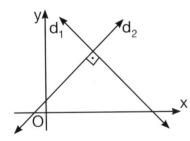

– Slopes are negative reciprocals of each other.

$$\left(d_1 \perp d_2, \text{then } m_1 = -\frac{1}{m_2} \right)$$

SOLVING EQUATIONS INVOLVING PARALLEL AND PERPENDICULAR LINES EXERCISES

Determinate whether the following pairs of lines are parallel, perpendicular, or neither.

1) $2x - 3y = 12$

$y = \dfrac{3x}{12} + 12$

2) $\dfrac{2x}{3} + y = 6$

$3x - 2y = 18$

3) $3x - 4y = 12$

$y = \dfrac{-4x}{3} + 12$

4) $x + y = 6$

$x = -y$

5) $\dfrac{1}{2}x + \dfrac{1}{4}y = 12$

$2x + y = 18$

6) $3x - 6y = 6$

$y = -2x$

Find an equation of the line that is **parallel** to the graph of each equation and passes through given point.

7) $y = 4x + 12;\ (2, 3)$

8) $y = \dfrac{1}{2}x - 6;\ (1, 5)$

9) $y = 5x - 7;\ (-1, -4)$

10) $y = \dfrac{2}{3}x - 18;\ (1, 3)$

Find an equation of the line that is **perpendicular** to the graph of each equation and passes through the given point.

11) $y = -x - 3;\ (2, -3)$

12) $\dfrac{1}{2}x - \dfrac{1}{3}y = 0;\ (-1, -4)$

13) $y + x = 9;\ (4, 8)$

14) $x = 2y - 5;\ (-1, -2)$

Key Notes:

Substitution Method:

Example: Solve the following system of equations using the **substitution method.**

$x - 2y = 18$

$2x + 3y = 15$

Solution:

Step 1: Solve for one variable of the equations (either y or x).

If $x - 2y = 18$ then $x = 2y + 18$

Step 2: Substitute this expression into the other equation.

$2x + 3y = 15$

$2(2y+18) + 3y = 15$

Step 3: Solve the equation for y.

$2(2y+18) + 3y = 15$

Then,

$4y + 36 + 3y = 15$

$7y = 15 - 36$

$7y = -21$

$y = -3$

Step 4: Substitute –3 for y in the equation.

$x = 2y + 18$

$x = 2(-3) + 18$

$x = -6 + 18$

$x = 12$

Solution (12, –3)

Elimination Method:

Example: Solve the following system of equations using the **elimination method.**

$3x - 2y = 24$

$2x + y = -5$

Step 1: Multiply the second equation by 2 and keep the first equation the same.

$2(2x + y) = -5(2)$ then...

$4x + 2y = -10$

Step 2: Add the revised equations and solve for x.

$$\begin{array}{r} 3x - 2y = 24 \\ + \quad 4x + 2y = -10 \\ \hline 7x = 14 \text{ then } x = 2 \end{array}$$

Step 3: Substitute the value of x into one of the equations and then solve for y.

$3x - 2y = 24$ if $x = 2$ then

$3(2) - 2y = 24$

$6 - 2y = 24$

$-2y = 24 - 6$

$-2y = 18$

$y = -9$

Solution $(2, -9)$.

SOLVING SYSTEMS OF EQUATIONS
BY SUBSTITUTION & ELIMINATION EXERCISES

Solve each system by Elimination

1) $2x + y = 20$

$3x - 4y = 19$

2) $x + 2y = 15$

$3x - 2y = 17$

3) $3x + 2y = 10$

$x + y = 14$

4) $2x + 6y = 18$

$2x - y = 4$

5) $x - \dfrac{1}{2}y = 16$

$x + \dfrac{1}{2}y = 6$

6) $7x - y = 118$

$2x - y = 8$

Solve each system by Substitution

7) $x = 2y + 2$

$x - 4y = 8$

8) $x + 2y = 18$

$y = x + 6$

9) $x = 3y$

$x + y = 20$

10) $2x - 3y = 0$

$2x + y = 12$

11) $x - \dfrac{1}{2}y = 25$

$x = 3y$

12) $\dfrac{1}{2}x = 3y$

$x + 3y = 36$

Solve the following equations.

13) $3k - 5 + 4k = -4 (k + 4)$

14) $\dfrac{5}{8} (8x + 24) = 4(x - 3)$

15) The perimeter of a rectangular garden is 48 cm. The width of the garden is 3 cm longer than 2 times the length. What is the length of the garden?

16) The cost of 4 child movie tickets and 2 adults movie tickets is $48. If the price of each child's ticket is one–third the price of each adult's ticket, what is the cost for 1 child ticket?

AMERICAN MATH
ACADEMY

FACTORING QUADRATIC EQUATIONS

Special Cases:

- $a^2 - b^2 = (a - b)(a + b)$

- $a^2 + b^2 = (a + b)^2 - 2ab$

- $a^3 - b^3 = (a - b)(a^2 + 2ab + b^2)$

- $a^3 + b^3 = (a - b)^3 + 3ab(a - b)$

- $a^3 + b^3 = (a + b)(a^2 - 2ab + b^2)$

- $a^3 + b^3 = (a + b)^3 - 3ab(a + b)$

- $\dfrac{a^2}{b^2} + \dfrac{b^2}{a^2} = \left(\dfrac{a}{b} + \dfrac{b}{a}\right)^2 - 2$

Foil Method:

Examples:

1. $x^2 + 5x + 4 = (x + 5)(x + 1)$

2. $x^2 - 7x + 12 = (x - 4)(x - 3)$

3. $x^2 - 25 = (x - 5)(x + 5)$

Factor each of the following equations.

1) $x^2 - 49$

2) $x^2 - 100$

3) $x^2 - 8x + 15$

4) $x^2 - x + 42$

5) $6x^2 - 21x - 12$

6) $x^2 - 2$

7) $\dfrac{x^2}{y^2} + \dfrac{y^2}{x^2}$

8) $x^2 + 2xy + y^2$

9) $x^2 - 6x - 16$

10) $x^2 - 9x + 8$

11) $7x^2 + 55x - 72$

12) $2x^2 - 7x - 15$

13) $25x^2 + 5x - 2$

14) $x^2 - x - 3$

15) For what values of k is the following expression factorable?

$x^2 + kx + 3$

SOLVING QUADRATIC EQUATIONS BY FORMULA
AND COMPLETE SQUARE

Key Notes:

Quadratic Equation: $ax^2 + bx + c$

Quadratic Formula: $x = \dfrac{-b \pm \sqrt{b^2 - 4ac}}{2a}$

Discriminant: $b^2 - 4ac$

- If $b^2 - 4ac < 0$, there are no real roots.

- If $b^2 - 4ac = 0$, there is 1 real root.

- If $b^2 - 4ac > 0$, there are 2 real roots.

Complete Square Method:

- Step 1: Write the equation in the form $ax^2 + bx = c$

- Step 2: Square half the coefficient of x, and add the result to both sides of the equation

- Step 3: Complete the square

- Step 4: Factor and simplify

- Step 5: Take the square root of both sides and solve equation

Example: $x^2 + 8x + 12 = 0$

- Step 1: $x^2 + 8x = -12$

- Step 2: $\dfrac{8}{2} = 4^2 = 16$

- Step 3: $x^2 + 8x + 16 = -12 + 16$

- Step 4: $(x+4)^2 = 4$

- Step 5: $(\sqrt{x+4})^2 = (\sqrt{4})^2$, then $x + 4 = \pm 2$, $x + 4 = 2$ or $x + 4 = -2$

$$x = -2 \text{ or } x = -6$$

SOLVING QUADRATIC EQUATIONS BY FORMULA AND COMPLETE SQUARE EXERCISES

Solve each of following quadratic equations by using the quadratic formula.

$$x = \frac{-b \pm \sqrt{b^2 - 4ac}}{2a}$$

1) $x^2 + 8x + 12 = 0$

2) $x^2 + 2x + 1 = 0$

3) $4x^2 + 5x + 1 = 0$

4) $x^2 - 8x + 12 = 0$

5) $x^2 - x - 1 = 0$

6) $x^2 - 4x + 3 = 0$

7) $x^2 - 6x + 1 = 0$

8) $3x^2 + 4x - 3 = 0$

Solve each of the following equations by using the complete square method.

9) $x^2 - 6x - 7 = 0$

10) $x^2 + 6x - 16 = 0$

11) $x^2 + 12x + 8 = 0$

12) $x^2 - 6x - 20 = 0$

13) $2x^2 + 8x - 10 = 0$

14) $3x^2 + 6x - 18 = 0$

15) $5x^2 + 20x - 40 = 0$

16) $x^2 + 4x + 3 = 0$

AMERICAN MATH
━ ACADEMY ━

Key Notes:

$$\begin{array}{c|c} P(x) & Q(x) \\ & B(x) \\ \hline \underline{-} & \\ K(x) & \end{array}$$

$$P(x) = Q(x) \cdot B(x) + K(x)$$

Monomial: 2x, 3x, ...etc.

Binomial: 2x + 4y, ...etc.

Trinomial: 2x + 3y + 5.

Not polynomials

Example:

$$3x + \frac{1}{x}$$

$$\sqrt{x} - 7x^2 + 7$$

Example: If P (x) = 2x − 3 and Q (x) = 5x − 4 find P (x) + Q (x) = ?

Solution: P (x) + Q (x) = (2x − 3) + (5x − 4) = 7x − 7

Example: If P (x) = 7x + 3 and Q (x) = 6x − 8 find P (x) − Q (x) = ?

Solution: P (x) − Q (x) = (7x + 3) − (6x − 8) = x + 11

ADDING AND SUBTRACTING POLYNOMIALS EXERCISES

Simplify each of following polynomials.

1) $(5x^2 + 12x) + (12x - 9)$

2) $(-7x^2 + 10x) + (8x - 11)$

3) $(-x^2 + 12x + 8) + (5x - 8)$

4) $(6x^2 + x) + (-8x - 13)$

5) $(12x^2 - 12x) + (-12x - 4)$

6) $(2x^2 + 7x + 4) + (2x - 18)$

7) $(x^2 + 8x + 6) - (2x - 18)$

8) $(-x^2 - 10x) - (-5x - 13)$

9) $(x^2 - x) - (-x^2 - 8x + 9)$

10) $(4x^2 - 3x) - (8x^2 - 8x)$

11) $(x^2 - 5)(x^2 - 6x - 12)$

12) $(7x^2 - x) - (x^2 - 14x + 4)$

13) $(x^2 - x)(-x^2 - 8x + 9)$

14) $(4x^2 - 3x) - (8x^2 + 8x)$

15) $(-5x^2 - 18x + 9) + (-x^2 + 8x + 4)$

16) $(4x^2 - 3x) - (-8x)$

MULTIPLYING AND DIVIDING POLYNOMIALS

Key Notes:

- $a^n \cdot a^m = a^{n+m}$

- $\dfrac{a^n}{a^m} = a^{n-m}$

- $a^0 = 1$

- $a^2 - b^2 = (a-b)(a+b)$

Foil Method:

Example: If $(a^2b)^3 \cdot (a^3 b^2) = ?$

Solution: $(a^2b)^3 \cdot (a^3b^2) = (a^6b^3) \cdot (a^3b^2) = a^9b^5$

Example: If $\dfrac{x^2-16}{x+4} = ?$

Solution: $\dfrac{x^2-16}{x+4} = \dfrac{(x-4)(x+4)}{x+4} = x-4$

MULTIPLYING AND DIVIDING POLYNOMIALS EXERCISES

Simplify each of following polynomials.

1) $(a^{12} b^4) \cdot (a^1 b^5)$

2) $(a^3 b^7) \cdot (a^8 b^{10})$

3) $(a^3)^4 \cdot (a^6 b^3)$

4) $(6a^3) \cdot (a^4 b^8)$

5) $(a^3 b^2)^4 \cdot (a^1 b^{15})$

6) $(a^{13} b^{17}) \cdot (a^8 b^{10})$

7) $(a^3 - 3a^2)(2a^3 - a^2)$

8) $(a^2 + a + 2) + (a^2 + 5a)$

9) $17a^4(2a^6)$

10) $3a^2(a^3 - a + 7)$

11) $\dfrac{15a - 25}{5}$

12) $\dfrac{6a^3 + 12a^2 - 60}{6}$

13) $\dfrac{a^4 + 5a^3 + 9a}{a}$

14) $3(x - 4)(2x + 1)$

15) $\left(\dfrac{a^2 - b^2}{a^2 + ab}\right)\left(\dfrac{a^2 + ab}{ab + a}\right)$

16) $\left(\dfrac{a^2 - b^2}{a^2 + ab}\right)\left(\dfrac{a^2 + ab}{ab + a^2}\right)$

AMERICAN MATH
ACADEMY

SOLVING EQUATIONS WITH ALGEBRAIC FRACTIONS

Key Notes:

Equation: A sentence that states that two mathematical expressions are equal.

Variable: Represents an unknown number. Any letter can be used as a variable.

Example: x, y, z, ... etc.

Coefficient: A number placed before a formula in an equation.

Example: ③x, ⑤(x + 5)

Constant: A symbol, which represents a specific number.

Example: ⑤, 3x + ⑩

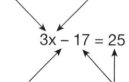

Coefficient Variable

$$3x - 17 = 25$$

Operator   Constants

Example: $\dfrac{x+3}{5} = \dfrac{2x+7}{4}$, Solve for x.

Solution: $\dfrac{x+3}{5} = \dfrac{2x+7}{4}$ (Cross multiply)

$4(4x + 3) = 5(2x + 7)$ (Distribute)

$16x + 12 = 10x + 35$ (Subtract 10x from both sides)

$16x - 10x + 12 = 10x - 10x + 35$

$4x + 12 = 35$ (Subtract 12 from both sides)

$4x + 12 - 12 = 35 - 12$

$4x = 23$ (Divide both sides by 4)

$x = \dfrac{23}{4}$

Solve each of the following equations.

1) $\dfrac{2x-3}{4} = \dfrac{x}{5}$

2) $\dfrac{x-3}{x-5} = \dfrac{2}{5}$

3) $\dfrac{x-1}{4} + \dfrac{x}{3} = \dfrac{1}{2}$

4) $\dfrac{2x+5}{6} + \dfrac{x}{18} = \dfrac{1}{9}$

5) $\dfrac{x}{4} + \dfrac{x}{8} = 7$

6) $\dfrac{x}{6} - \dfrac{x}{24} = 48$

7) $\dfrac{x-1}{4} + \dfrac{x}{8} = \dfrac{7x}{2}$

8) $\dfrac{x-1}{2} + \dfrac{x+1}{14} = 5$

9) $\dfrac{x}{4} - 5 = \dfrac{x}{2}$

10) $\dfrac{x}{6} + 7 = \dfrac{x}{3}$

11) $\dfrac{x-12}{4} = \dfrac{x-9}{2}$

12) $\dfrac{2x+5}{36} = \dfrac{1}{6}$

13) $\dfrac{\frac{1}{x}}{\frac{3}{x-1}} = \dfrac{1}{4}$

14) $\dfrac{1}{x} + \dfrac{1}{12} = \dfrac{1}{36}$

15) $\dfrac{\frac{x-2}{3}}{\frac{x-5}{4}} = \dfrac{1}{12}$

16) $\dfrac{x^2-4}{x-2} = 6$

17) $\dfrac{x^2-36}{x-6} = 12$

18) $\dfrac{1}{x} + \dfrac{1}{2x} = \dfrac{1}{36}$

SIMPLIFYING RATIONAL EXPRESSIONS

Key Notes:

- **A rational expression** is a fraction that can be written as a ratio of two polynomials.
- When the denominator is zero, the rational expression is **undefined**.

If two fractions have unlike denominators:

- Find the smallest multiple (LCD) of both numbers.
- Rewrite the fractions as equivalent fractions with the LCD as the denominators.

Example: Simplify $\dfrac{x^2-4}{x+2} \cdot \dfrac{x^2-9}{x+3}$

Solution: $\dfrac{x^2-4}{x+2} \cdot \dfrac{x^2-9}{x+3}$

$\dfrac{(x-2)(x+2)}{x+2} \cdot \dfrac{(x-3)(x+3)}{x+3}$

$\dfrac{x-2}{1} \cdot \dfrac{x-3}{1}$

$(x-2)(x-3) = x^2 - 5x + 6$

Example: Simplify $\dfrac{1}{x+1} + \dfrac{1}{x-1}$

Solution: $\dfrac{1}{x+1}\left(\dfrac{x-1}{x-1}\right) + \dfrac{1}{x-1}\left(\dfrac{x+1}{x+1}\right)$

$\dfrac{x-1}{x^2-1} + \dfrac{x+1}{x^2-1}$

$\dfrac{x-1+x+1}{x^2-1}$

$\dfrac{2x}{x^2-1}$

Find the sum or difference of each of following.

1) $\dfrac{1}{x+2} + \dfrac{1}{x-2}$

2) $\dfrac{1}{-x-4} - \dfrac{1}{x+4}$

3) $\dfrac{x}{x-1} + \dfrac{2x}{x+1}$

4) $\dfrac{1}{3} - \dfrac{1}{x+4}$

5) $\dfrac{2x}{x-1} - \dfrac{5}{x+3}$

6) $\dfrac{1}{3x} + \dfrac{1}{x-1}$

Simplify each of following.

7) $\dfrac{2x^2}{12x^6}$

8) $\dfrac{3x^{-4}}{9x^8}$

9) $\dfrac{x^2-36}{x-6}$

10) $\dfrac{3x^{-4}y^{-7}}{9x^{12}y^{-5}}$

11) $\dfrac{x(x+3)}{3x}$

12) $\dfrac{x^{-1}y^{-3}z^4}{x^{10}y^4z^{-2}}$

Multiply or divide each of following.

13) $\dfrac{x^2-y^2}{3x} \cdot \dfrac{9x}{x-y}$

14) $\dfrac{x^2}{3x} \cdot \dfrac{9x}{36}$

15) $\dfrac{x^2-y^2}{x} \div \dfrac{x+y}{x}$

16) $\dfrac{x^2}{3} \cdot \dfrac{18}{x^8}$

17) $\dfrac{5x(x-1)}{3x} \cdot \dfrac{1}{x-1}$

18) $\dfrac{x^3-y^3}{x^2+xy+y^2}$

FUNCTIONS NOTATION AND INVERSE FUNCTION

Key Notes:

Function: If every element of A is paired with the elements of B at least once and A≠0 and B≠0, this correlation is called a function.

Inverse Function: Let f be a one to one function with domain of x and range of y. The inverse of this function is $f^{-1}\{y\} = x$.

$f(x) = y \Leftrightarrow f^{-1}(y) = x$

Function Notation

Name of Function

Output

Input

$y = F(x)$

Example:

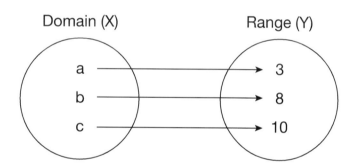

Not a Function: If the domain (x) is repeating, it's not a function. You do not need to check the range (y). It does not matter if the range repeats.

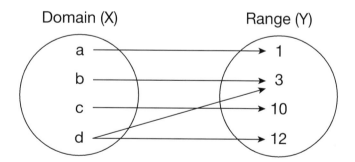

Examples:

(a, 1), (b, 3), (c, 5) is a function because domain elements are not repeating.

(a, 1), (b, 3), (c, 5),(a,10) is not a function because domain elements are repeating.

FUNCTIONS NOTATION AND INVERSE FUNCTION EXERCISES

For each of following points, state whether or not it is a function.

1) (2, 3), (5, 3),(6, 7)

2) (1, 3), (2, 4),(3, 5), (1,7)

For each of following tables state whether the relation represents a function.

3)

x	F(x)
e	1
l	3
m	1

4)

x	F(x)
a	1
b	3
c	5

5)

x	F(x)
m	4
n	5
k	4

6)

x	F(x)
John	Smart
Peter	Intelligent
Peter	Genius

Find the Inverse of each of the following functions.

7) $f(x) = x + 4$

8) $f(x) = 2x - 54$

9) $f(x) = x^2 - 3$

10) $f(x) = \sqrt{x - 6}$

Key Notes:

Function Operations

1) Adding: $(f + g)(x) = f(x) + g(x)$

Example: Let $f(x) = 4x-2$ and $g(x) = 2x - 1$. Then $(4x - 2) + (2x - 1) = 6x - 3$

2) Subtracting: $(f - g)(x) = f(x) - g(x)$

Example: Let $f(x) = 4x - 2$ and $g(x) = 2x - 1$, then $(4x - 2) - (2x - 1) = 2x - 1$

3) Multiplying: $(f \cdot g)(x) = f(x) \cdot g(x)$

Example: Let $f(x) = 4x - 2$ and $g(x) = 2x - 1$, then $(4x - 2).(2x - 1) = 8x^2 - 8x + 2$

4) Dividing: $(f \div g)(x) = f(x) \div g(x)$

Example: Let $f(x) = 4x - 2$ and $g(x) = 2x - 1$,

$$\text{then } (4x - 2) \div (2x - 1) = \frac{4x - 2}{2x - 1} = \frac{2(2x - 1)}{2x - 1} = 2$$

FUNCTION OPERATIONS EXERCISES

Adding and Subtracting Functions.

Let $f(x) = -3x + 9$ and $g(x) = 2x + 5$

1) Find $f(x) + g(x)$

2) Find $f(x) - g(x)$

Let $f(x) = x^2 + 3x + 9$ and $g(x) = x + 12$

3) Find $f(x) + 2g(x)$

4) Find $3f(x) - g(x)$

Find each value or expression

Let $f(x) = x + 4$ and $g(x) = 3x + 10$

5) Find $f(g(x))$

6) Find $g(f(x))$

Let $f(x) = 5x$ and $g(x) = 6x - 15$

7) Find $f(g(x))$

8) Find $g(f(x))$

Let $f(x) = 2x^2 + 1$ and $g(x) = x - 3$

9) Find $f(g(3))$

10) Find $g(f(5))$

Let $f(x) = 2x^2 + x + 3$ and $g(x) = 4x^2 + 5x - 12$

11) Find $f(g(-1))$

12) Find $g(f(-3))$

Multiplying and Dividing Functions.

Let $f(x) = x^2 - x$ and $g(x) = x^2 + x - 2$

13) Find $f(x) \cdot g(x)$

14) Find $\dfrac{f(x)}{g(x)}$

MEAN, MEDIAN AND MODE

Key Notes:

Mean (Average): The sum of the numbers divided by how many numbers there are.

$$\text{Mean} = \frac{\text{Sum of the numbers}}{\text{How many numbers}}$$

Example: Find the mean of 3, 5, 6, 7, and 9.

Solution:

$$\text{Mean} = \frac{\text{Sum of the numbers}}{\text{How many numbers}}$$

$$\text{Mean} = \frac{3+5+6+7+9}{\text{How many numbers}} = \frac{30}{5} = 6$$

Median: The middle number.

- Put all of the numbers in order, from smallest to largest
- The median is the middle number.
- If there is an even amount of numbers, the median is the average of the two middle numbers.

Example: Find the median of 3, 12, 20, 8, 16, and 30.

Solution: Put all of the numbers in order: 3, 8, 12, 16, 20, 30. The middle numbers are 12 and 16.

The averag of these two number is $\frac{12+16}{2} = 14$

Mode: The most frequent value.

Example: Final the mode: A, B, C, D, C, A, B, A, C, D, C

Solution: Most repeated letter is C

Range: The difference between the lowest and highest value.

Example: Find the range of 2, 4, 20, 30, 50, and 80.

Solution: Range 80 − 2 = 78

MEAN, MEDIAN AND MODE EXERCISE

Find the mean, median, mode, and range for each set of numbers.

1. 3, 4, 5, 6, 6, 7, 11

Mean:

Median:

Mode:

Range:

2. 1, 3, 5, 7, 7, 7

Mean:

Median:

Mode:

Range:

3. $\dfrac{1}{2}, \dfrac{1}{3}, \dfrac{1}{4}, \dfrac{1}{5}$

Mean:

Median:

Mode:

Range:

Find the missing number using data points and mean.

4. The average of 6
and y is 10

y: _____

5. The average of 12
and z is 20

z: _____

6. The average of 12.5
and k is 30.5

k: _____

7. The average of five consecutive positive integers is 30. What is the greatest possible value of one of these integers?

8. What measure of central tendency is calculated as the difference between the lowest and highest the number of values?

SLOPE AND SLOPE INTERCEPT FORM

Key Notes:

Slope Intercept Form:

$y = mx + b$ m : slope

b : y – intercept $= (0, b)$

Suppose there are two points on a line, (x_1, y_1) and (x_2, y_2).

The slope m of the line is:

$$m = \frac{\text{change in y (Rise)}}{\text{change in x (Run)}} = \frac{y_2 - y_1}{x_2 - x_1}$$

Example:

Find the equation of the line that passes through the points (1, 2) and (3, 6)

Solution:

$$\text{Slope} = m = \frac{6-2}{3-1} = \frac{4}{2} = 2$$

Example: Find slope of $y = 3x - 7$

Solution: If $y = mx + b$, then $m = 3$

Example: Find slope of $y = \frac{1}{7}x + 13$

Solution: The equation is in slope intercept form ($y = mx + b$), so slope $= m = \frac{1}{7}$

SLOPE AND SLOPE INTERCEPT
FORM EXERCISE

Find the slope of the line that passes through each pair of points. $\left[m = \dfrac{y_2 - y_1}{x_2 - x_1} \right]$

1) (3, 4) and (7, 9)

2) (2, 1) and (6, 8)

3) (4, 9) and (3, 7)

4) (–3, 6) and (–1, –7)

5) (3, 2) and (1,8)

6) (0, 3) and (3, 2)

7) $\left(-2, \dfrac{1}{3}\right)$ and $\left(1, \dfrac{-1}{3}\right)$

8) $\left(2, \dfrac{1}{2}\right)$ and $\left(6, \dfrac{1}{4}\right)$

Find the equation of the line in slope–intercept form. [y = mx + b]

9) m = 3 and y –intercept = –19

10) m = 6 and (–1, 3)

11) m = 5 and y– intercept = 6

12) m = –2 and (–4, –6)

13) m = $\dfrac{-1}{3}$ and y–intercept 6

14) m = $\dfrac{-2}{5}$ and y–intercept $\dfrac{1}{5}$

Find the slope of each of following equations.

15) y = $\dfrac{1}{4}$x + 5

16) y = $\dfrac{-1}{5}$x + 10

17) y = –7x + 12

18) y = 8x – 18

AMERICAN MATH
ACADEMY

Key Notes:

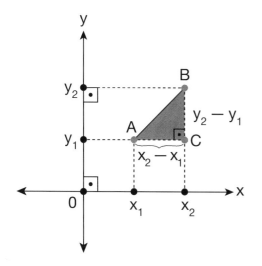

Use the Pythagorean theorem to find the length of $\overline{AB}$, which is the hypotenuse of the right triangle.

$$|AB|^2 = (x_2 - x_1)^2 + (y_2 - y_1)^2$$

$$|AB| = \sqrt{(x_2 - x_1)^2 + (y_2 - y_1)^2}$$

Midpoint formula:

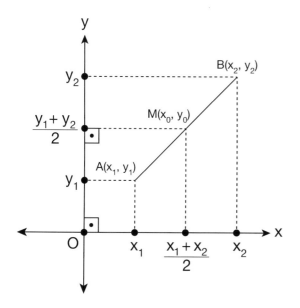

What are coordinates (X_0, Y_0) of the midpoint of the segment whose endpoints are $A(x_1, y_1)$ and $B(x_2, y_2)$?

$$x_0 = \frac{x_1 + x_2}{2}$$

$$y_0 = \frac{y_1 + y_2}{2}$$

DISTANCE AND MIDPOINT EXERCISE

Use the distance formula to find the distance between each pair of points.

Distance formula: $d = \sqrt{(x_2 - x_1)^2 + (y_2 - y_1)^2}$

1) (3, 5) and (7, 8)

2) (−1, 3) and (−5, −8)

3) (6, 4) and (−3, 12)

4) $\left(\frac{1}{2}, 6\right)$ and $\left(\frac{1}{3}, 6\right)$

5) $\left(\frac{1}{2}, 2\right)$ and $\left(\frac{1}{2}, 8\right)$

6) $\left(\frac{1}{2}, 0\right)$ and $\left(-\frac{1}{4}, 0\right)$

7) (6, 7) and (7, 3)

8) (10, 14) and (2, 8)

9) (6, 8) and (−1, −3)

10) (9, −3) and (−4, −1)

Find the midpoint of the given points. Mid point formula: $M = \left(\dfrac{x_1 + x_2}{2}, \dfrac{y_1 + y_2}{2}\right)$

11) (0, 5) and (2, 9)

12) (−2, 4) and (−8, −8)

13) (7, 6) and (−1, 0)

14) $\left(\frac{1}{3}, 9\right)$ and $\left(\frac{1}{3}, 9\right)$

15) $\left(\frac{1}{6}, 1\right)$ and $\left(-\frac{1}{6}, 7\right)$

16) $\left(\frac{1}{8}, 0\right)$ and $\left(\frac{1}{8}, 0\right)$

17) (9, 7) and (7, 3)

18) (−1, −13) and (9, 23)

AMERICAN MATH
ACADEMY

Key Notes:

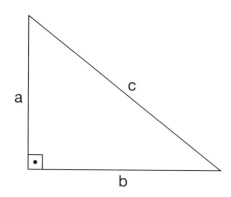

$$a^2 + b^2 = c^2$$

Example: What ise the value of x?

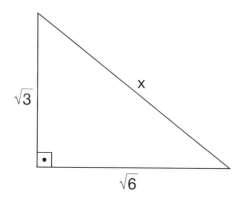

Solution:

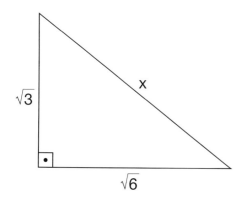

$$(\sqrt{3})^2 + (\sqrt{6})^2 = x^2$$

$$3 + 6 = x^2$$

$$9 = x^2$$

$$3 = x$$

PYTHAGOREAN THEOREM EXERCISE

For each of the following triangles find the missing length.

1)

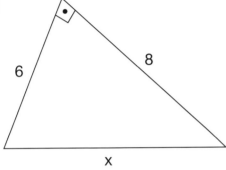

2)

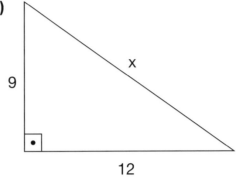

3)

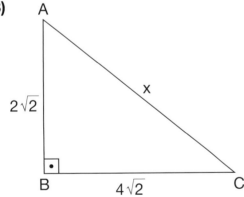

4)

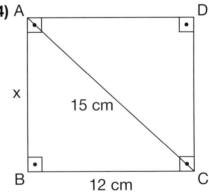

AMERICAN MATH
ACADEMY

5)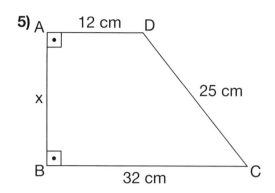

6)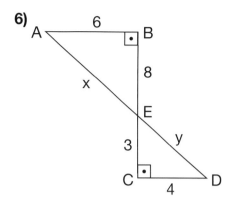

7)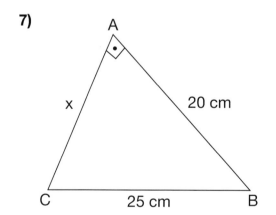

8)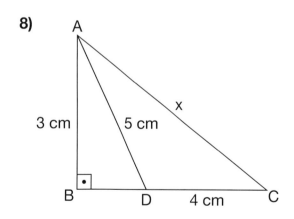

1. If $27^x = 81^y$, what is the ratio of y to x?

A) $\dfrac{3}{2}$

B) $\dfrac{3}{4}$

C) 4

D) 3

2. If the sum of three consecutive integers is 96, which of the following is the largest integer?

A) 31

B) 32

C) 33

D) 34

3. What is the solution to the equation below?

$2(y - 6) = 3(2y - 12)$

A) –6

B) 6

C) 1

D) $\dfrac{1}{6}$

4. For what value of x is $\dfrac{x+4}{3} - \dfrac{x}{5} = \dfrac{14}{15}$?

A) –1

B) –2

C) –3

D) –4

5. If $x = \dfrac{5^3}{\sqrt{125}}$, then x = ?

A) $5\sqrt{5}$

B) $2\sqrt{5}$

C) $3\sqrt{5}$

D) $4\sqrt{5}$

6. Simplify the following equation.

$\dfrac{x^2 - 25}{x + 5}$

A) $\dfrac{x-5}{x+5}$

B) $\dfrac{x+3}{x-3}$

C) $x + 5$

D) $x - 5$

American Math Academy

7. Simplify $\sqrt{32} - \sqrt{8} + \sqrt{128}$

A) $-10\sqrt{2}$

B) 10

C) $\sqrt{2}$

D) $10\sqrt{2}$

8. Simplify $\dfrac{x^2 - y^2}{x^2 - xy} \cdot \dfrac{xy - x}{x^2 + xy}$

A) $x + y$

B) $y + 1$

C) $\dfrac{y - 1}{x}$

D) $\dfrac{y + 1}{x}$

9. What is the vertex of the parabola $y = 3x^2 + 6x - 9$?

A) $f(x) = 3(x + 1)^2 - 12$

B) $f(x) = 3(x + 1)^2 - 6$

C) $f(x) = -3(x - 1)^2 - 4$

D) $f(x) = -3(x - 1)^2 - 9$

10. What is simplest form of $\sqrt{-8} \cdot \sqrt{32}$?

A) $4i$

B) $8i$

C) $16i$

D) $4i\sqrt{2}$

11. Which of the following expressions in the form a+bi is equivalent to $\dfrac{3+i}{2-i}$?
(Note $i^2 = -1$)

A) $1 + i$

B) i

C) 1

D) $1 - i$

12. If $2x - 18 = 5x - 14$, what is the value of x?

A) 3

B) 4

C) $\dfrac{3}{4}$

D) $-\dfrac{4}{3}$

American Math Academy

MIXED REVIEW TEST III

13. If $\dfrac{a^{2m}}{a^{10}} = a^6$ and $a^{3n} = a^{30}$, then what is the value of $m \cdot n$?

A) 50

B) 60

C) 80

D) 90

14. If $2^a \cdot 2^a \cdot 2^a \cdot 2^a = 8^{4b}$, find a in terms of b?

A) 3b

B) −3b

C) 4b

D) 8b

15. What is the product of $(3 + 4i) \cdot (3 - 4i)$? $(i = \sqrt{-1})$

A) 9

B) 16

C) 25i

D) 25

16. Solve for x.

$$x^2 - 4x + y = 14$$
$$5x - y = 6$$

A) (−1, 4)

B) (4, 5)

C) (−4, −5)

D) (4, −5)

17. What are the zeros of $f(x) = x^2 - 169$?

A) x = −13

B) x = ± 13

C) x = ± 13i

D) x = 13i

American Math Academy

1. If $\dfrac{1}{2^{3x}} = \dfrac{1}{32^2}$, find x.

A) $\dfrac{1}{3}$

B) $\dfrac{3}{10}$

C) 10

D) $\dfrac{10}{3}$

2. If y is directly proportional to x and y is equal to 18, when x is equal to 3, what is the value of y when x is 4?

A) 12

B) 24

C) 36

D) 72

3. If $\dfrac{7}{a} = \dfrac{4}{b}$, what is the value of $\dfrac{a}{b}$?

A) $\dfrac{7}{4}$

B) $\dfrac{4}{7}$

C) 4

D) 7

4. If $\dfrac{1}{3}(3x - 9) + (x - 12) = ax + x + b$, what is the value of a − b?

A) 15

B) 16

C) 17

D) 18

5. "A line is represented by equation $\dfrac{3}{5}x - ny = 25$." If the slope of the equation is $\dfrac{1}{25}$, what is the value of n?

A) 3

B) 5

C) 12

D) 15

6. Line K passes through the coordinates (4, 8) and (10,14). What is the slope of line K?

A) 1

B) 2

C) 3

D) 4

7. If $3x + 7 = -2x - 8$, what is the value of x?

A) 2

B) 3

C –3

D) –5

8. The price of a book has been discounted 20%. The sale price is $60. What is the original price?

A) $25

B) $45

C) $65

D) $75

9. If $x = \dfrac{2^3}{\sqrt{12}}$, What is the value of x?

A) $4\sqrt{3}$

B) $\sqrt{3}$

C) $\dfrac{4\sqrt{3}}{3}$

D) $\dfrac{3\sqrt{3}}{4}$

10. If $9^{2x-2} = 81^{2x-3}$, then what is the value of x?

A) 1

B) –1

C) –2

D) 2

11. $x^2 - 8x - 12 = 0$, solve for x.

A) $x = 4 \pm 2\sqrt{7}$

B) $x = -4 \pm 2\sqrt{7}$

C) $x = \sqrt{7} \pm 4$

D) $x = \sqrt{7}$

12. $$i^{2020} + i^{2021} + i^{2022}$$

Which of the following is equivalent to the complex number shown above?

A) $1 - i$

B) $1 + i$

C) i

D) 1

American Math Academy

13. If $\dfrac{3}{4}x - \dfrac{1}{8}x = \dfrac{1}{12} + \dfrac{2}{3}$, what is the value of x?

A) 6

B) 7

C) $\dfrac{6}{7}$

D) $\dfrac{6}{5}$

14. Simplify the following equation.

$$\dfrac{x^4 - 16}{x^2 - 4} = ?$$

A) $\dfrac{x - 4}{x + 4}$

B) $\dfrac{x^2 + 4}{x^2 - 2}$

C) $x^2 - 4$

D) $x^2 + 4$

15. If $a = 2\sqrt{3}$ and $3a = \sqrt{3x}$, what is the value of x?

A) 18

B) 24

C) 36

D) 72

16. If the area of a circle is 16π, what is the circumference of the circle?

A) 4π

B) 8π

C) 10π

D) 12π

17. What are the zeros of the function $f(x) = x^3 + 6x^2 + 9x$?

A) 0, 2

B) 0, –2

C) 0, 3

D) 0, –3

American Math Academy

1. $\dfrac{3(x+5)-8}{7} = \dfrac{17-(6-x)}{5}$

In the equation above, what is the value of x?

A) $\dfrac{21}{4}$

B) $\dfrac{11}{4}$

C) $\dfrac{4}{21}$

D) $\dfrac{23}{4}$

2. If x is a positive number and $x^2 - 6x + 11 = 2$, what is the value of x?

A) 3

B) 4

C) 5

D) 6

3. The formula $F = \dfrac{9}{5}C + 32°$ shows the relation of temperature, in degrees Fahrenheit to degrees Celsius. Find F when C = 15°.

A) 15°

B) 25°

C) 42°

D) 59°

4. Simplify the following equation.

$\dfrac{x^2-16}{x-4}$

A) $\dfrac{x-4}{x+4}$

B) $\dfrac{x-2}{x+2}$

C) $x - 4$

D) $x + 4$

5. If $\dfrac{a^{2m}}{a^{10}} = a^{12}$ and $a^{3n} = a^{21}$, what is the value of $m \cdot n$?

A) 55

B) 66

C) 77

D) 88

6. What are the zeros of the function $f(x) = x^3 + 8x^2 + 16x$?

A) 0, –2

B) 0, –3

C) 0, –4

D) 0, –5

American Math Academy

7. $\dfrac{2x+4}{6} - \dfrac{x}{4} = \dfrac{3}{2}$, find the value of x.

A) 10

B) 12

C) 16

D) 18

8. Evaluate $4 + (2 + 3) \cdot 4 - 8 + (5 + 3)^0$

A) 15

B) 16

C) 17

D) 18

9. What is the solution of following equation?
$\dfrac{3}{x+1} = \dfrac{1}{x-1}$

A) 2

B) 3

C) 4

D) 5

10. Which of following lists all the factors of 48?

A) 1, 2, 3, 5, 7, 8, 12, 24, 16, 48

B) 1, 3, 4, 5, 8, 12, 16, 24, 48

C) 1, 2, 3, 4, 6, 8, 12, 16, 24, 48

D) 1, 2, 3, 4, 6, 8, 9, 12, 16, 24, 48

11. If x is the greatest prime factor of 14 and y is the greatest prime factor of 66, what is the value of x + y?

A) 17

B) 11

C) 18

D) 21

12. Which of following equations is an example of inverse variation?

A) $y = kx$

B) $y = \dfrac{k}{x}$

C) $y = \dfrac{x}{k}$

D) $y = x$

American Math Academy

13. A television originally priced at $400 is decreased in price by 25%. What is the sale price?

A) $225

B) $250

C) $275

D) $300

14. What measure of central tendency is calculated by the difference between the lowest and highest values?

A) Median

B) Mean

C) Mode

D) Range

15. The mean of five numbers is 48. If four of the numbers are 34, 54, 64 and 72 what is the value of the fifth number?

A) 4

B) 8

C) 12

D) 16

16. Simplify $\sqrt{(16)^2} + \sqrt{2^2} - (-2)^3$

A) 15

B) 18

C) 22

D) 26

17. If $4^{x-4} = 8^{x-6}$, then what is the value of x?

A) 5

B) 10

C) −5

D) −10

18. What is the value of k in the equation shown below?

$2k - 5 + 3k = -4(k + 4)$

A) $-\dfrac{11}{9}$

B) $\dfrac{11}{9}$

C) 9

D) 11

American Math Academy

19. What is the value of x in the equation shown below?

$$\frac{1}{2}(4x - 6) + 5 = 3x - 7$$

A) 3

B) 6

C) 9

D) 12

20. What is the greatest possible integer value of x?

$$\frac{1}{6} < 2x - 3 < \frac{3}{2}$$

A) 0

B) 1

C) 2

D) 3

American Math Academy

1. Solution:

Any real number is a number that cannot be written in fraction form or non–repeating decimals.

Correct Answer : C

2. Solution:

A rational number is a number that can be in the form B A and A is not equal to zero.

Correct Answer : D

3. Solution:

Reciprocal is the multiplicative inverse of a number

Correct Answer : B

4. Solution:

Subtract 36 from 48 = 48 – 36, then divided by three = (48 – 36) ÷ 3

Correct Answer : C

5. Solution:

$(4 \times 5) - (6 + 3) - 63 \div 7 = 20 - 9 - 63 \div 7 = 20 - 9 - 9 = 20 - 18 = 2$

Correct Answer : B

6. Solution:

Thirty–four is four times a number increased by nine.

$34 = 4x + 9$

Correct Answer : B

7. Solution:

$(5x^4 + 12x^2) - (12x^3 - 9x) = 5x^4 + 12x^2 - 12x^3 + 9x = 5x^4 - 12x^3 + 12x^2 + 9x$

Correct Answer : B

8. Solution:

Use prime factorization to find the greatest prime factor of 21 and 57.

If x is the greatest prime factor of 21, then $21 = 3.7$ and the greatest prime factor of 21 is 7. $(x = 7)$

If y is the greatest prime factor of 57, then $57 = 3.19$ and the greatest prime factor of 57 is 19. $(y = 19)$

$x + y = 7 + 19 = 26$

Correct Answer : C

9. Solution:

Let x equal the total number of teachers in school.

$$\left. \begin{array}{l} \text{if } 30 \text{ teachers is } 40\% \text{ of all} \\ x \text{ teachers is } 100\% \text{ of all} \end{array} \right\} \text{cross multiply}$$

$$x \cdot 40\% = 30 \cdot 100\%, \text{ then } x = \frac{30 \cdot 100}{40} = \frac{3000}{40}$$

$$= 75 \text{ teachers}$$

Correct Answer : D

American Math Academy

10. Solution:

$$\frac{x^2-8x+15}{x^2-9} \div \frac{x^2-4x-5}{x^2+3x}$$

$$\frac{(x-5)(x-3)}{(x-3)(x+3)} \cdot \frac{x(x+3)}{(x-5)(x+1)}$$

$$= \frac{x}{x+1}$$

Correct Answer : B

11. Solution:

Increasing tuition $660 \cdot \dfrac{15}{100} = \dfrac{9900}{100} = \$99$

Next year tuition total $= \$660 + \$99 = \$759$

Correct Answer : D

12. Solution:

$$18 = 2 \cdot 3^2$$
$$18^n = (2 \cdot 3^2)^n$$
$$= 2^n \cdot (3^2)^n$$
$$= 2^n \cdot (3^n)^2$$
$$= a^2 \cdot b$$

Correct Answer : B

13. Solution:

$$(81x^8)^{\frac{1}{4}} = (3^4 x^8)^{\frac{1}{4}} = 3x^2$$

Correct Answer : B

14. Solution:

$$\frac{\sqrt{a}}{2-\sqrt{a}} = \frac{\sqrt{a} \cdot (2+\sqrt{a})}{(2-\sqrt{a})(2+\sqrt{a})} = \frac{2\sqrt{a}+a}{4-a}$$

Correct Answer : A

15. Solution:

Ratio of students in science class to math class is 4 : 9.

If the number of students in math class is 27, then $9x = 27$ and $x = 3$

Number of students in science class is $4x = 4 \cdot 3 = 12$

Correct Answer : C

16. Solution:

Let $5x$ = smaller number,

$7x$ = larger number

$$7x = \frac{3}{5}(5x) + 20$$

$$7x = 3x + 20$$

$$7x - 3x = 20$$

$$4x = 20$$

$$x = 5$$

Larger number $= 7x = 7 \cdot 5 = 35$

Correct Answer : D

17. Solution:

$$\left(\frac{a}{b}\right) \cdot \left(\frac{ab}{cd}\right) \cdot \left(\frac{cd}{a^2}\right) = \frac{a^2 \cdot b \cdot c \cdot d}{a^2 \cdot b \cdot c \cdot d} = 1$$

Correct Answer : D

American Math Academy

18. Solution:

If x is a positive number and $x^2 + x - 12 = 0$,
then $(x - 3)(x + 4) = 0$

$x - 3 = 0$, or $x + 4 = 0$

$x = 3$ or $x = -4$

Correct Answer : B

19. Solution:

Since x and y are integers, x can be –7 for max, and y can be 10 for max

$x^2 + y^2 = (-7)^2 + (10)^2 = 49 + 100 = 149$

Correct Answer : C

20. Solution:

First plan: $23 membership fee and $7 per visit

$y = \$23 + 7x$

Second plan: only per visit fee of $13

$y = 13x$ (no membership fee)

Correct Answer : B

21. Solution:

Final grade: $0.6 \cdot 80 + 0.3 \cdot x$

$0.6 \cdot 80 + 0.3 \cdot x \geq 90$

Correct Answer : D

22. Solution:

The absolute value inequality is equivalent to $|2x - 7| > 9$ or $|2x - 7| < -9$.

First Inequality		Second Inequality
$2x - 7 > 9$	Write inequalities.	$2x - 7 < -9$
$2x > 16$	Add 7 to each side.	$2x < -2$
$x > 8$	Divide each side by 2.	$x < -1$

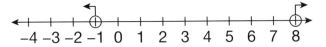

Correct Answer : A

23. Solution:

Since the population of a city increased from 270 thousand to 360 thousand

Percent of increase:

$$\frac{360 - 270}{270} = \frac{90}{270} = \frac{9}{27} = \frac{1}{3} = 33.\overline{3}\%$$

Correct Answer : C

24. Solution:

$3x + 7 > 10 \rightarrow$ Seven added to the product of 3 and x is greater than 10

Correct Answer : B

American Math Academy

25. Solution:

$$\frac{2x-8}{3} - \frac{x+4}{4} = \frac{3}{2}$$

$$\frac{4(2x-8)}{12} - \frac{3(x+4)}{12} = \frac{3 \cdot 6}{12}$$

$$\frac{8x-32}{12} - \frac{3x+4}{12} = \frac{18}{12} \text{(Cancel denominator)}$$

$$8x - 32 - (3x - 12) = 18$$

$$8x - 3x - 32 - 12 = 18$$

$$5x - 44 = 18$$

$$5x = 44 + 18$$

$$5x = 62$$

$$x = 12 \cdot 4$$

Correct Answer : C

26. Solution:

$$\frac{1}{\dfrac{1}{x+3} + \dfrac{1}{x+5}} =$$

$$\frac{1}{\dfrac{(x+5)}{(x+3)\cdot(x+5)} + \dfrac{(x+3)}{(x+5)\cdot(x+3)}} =$$

$$\frac{1}{\dfrac{2x+8}{x^2+5x+3x+15}} = \frac{x^2+8x+15}{2x+8}$$

Correct Answer : B

27. Solution:

$$\tan\alpha = \frac{2}{3}$$

(0, 0) is satisfied

$$y - 0 = \frac{2}{3}(x - 0)$$

$$y = \frac{2}{3}x$$

Correct Answer : B

28. Solution:

$$\frac{1}{x-1} + \frac{1}{x+1} = \frac{x+1+x-1}{(x-1)(x+1)}$$

$$= \frac{2x}{(x-1)(x+1)}$$

$$= \frac{2x}{x^2-1}$$

Correct Answer : D

29. Solution:

$$\frac{x^2-7x+12}{x^2-9x+20} = \frac{(x-3)\cdot(x-4)}{(x-4)\cdot(x-5)}$$

$$= \frac{x-3}{x-5}$$

Correct Answer : B

30. Solution:

$$\frac{1}{a} = \frac{1}{b} + \frac{1}{c}$$

$$\frac{1}{a} = \frac{c+b}{bc} \text{ (cross multiply)}$$

$$ac + ab = bc$$

$$ac = bc - ab$$

$$ac = b(c - a)$$

$$\frac{ac}{c - a} = b$$

Correct Answer : A

31. Solution:

$$\frac{18}{6} - 2 \div 4$$

$$= 3 - 2 \div 4$$

$$= 3 - \frac{1}{2}$$

$$= \frac{6 - 1}{2}$$

$$= \frac{5}{2}$$

Correct Answer : C

32. Solution:

$$= \sqrt{\frac{34 - 2}{2}} + 4^3 - 32$$

$$= \sqrt{\frac{32}{2}} + 64 - 32$$

$$= \sqrt{16} + 64 - 32$$

$$= 4 + 64 - 32$$

$$= 36$$

Correct Answer : D

33. Solution:

$$2a + a + 3b - 2b$$

$$= 3a + 3b - 2b$$

$$= 3a + b$$

Correct Answer : B

34. Solution:

$$2^2 a^3 + 4b^3 + a^3 - 2b^3$$

$$4a^3 + 4b^3 + a^3 - 2b^3$$

$$= 4a^3 + a^3 + 4b^3 - 2b^3$$

$$= 5a^3 + 2b^3$$

Correct Answer : A

35. Solution:

$$\sqrt{a^6} \cdot \sqrt{a^8}$$

$$= \sqrt{a^{8+6}}$$

$$= \sqrt{a^{14}}$$

$$= a^7$$

Correct Answer : C

36. Solution:

$$\sqrt{75}$$

$$= \sqrt{3 \cdot 25}$$

$$= 5\sqrt{3}$$

Correct Answer : B

37. Solution:

$\sqrt{21} + \sqrt{84}$

$= \sqrt{21} + \sqrt{21 \cdot 4}$

$= \sqrt{21} + 2\sqrt{21}$

$= 3\sqrt{21}$

Correct Answer : C

38. Solution:

$= (3x^6)^5 = 3^5 x^{30}$

$= 243x^{30}$

Correct Answer : A

39. Solution:

$\dfrac{(x^2y^4)^4}{x^7y^8}$

$= \dfrac{x^8y^{16}}{x^7y^8}$

$= x^{8-7}y^{16-8}$

$= x^1y^8$

Correct Answer : D

40. Solution:

The prime factorization of

$18 = 2 \cdot 3 \cdot 3 = 2 \cdot 3^2$

Correct Answer : B

41. Solution:

$\dfrac{4x - 2y}{2x - y}$

$= \dfrac{2\,\overset{1}{\cancel{(2x - y)}}}{\underset{1}{\cancel{2x - y}}}$

$= 2$

Correct Answer : B

42. Solution:

$\dfrac{\dfrac{3x^2}{4y^3}}{\dfrac{x^{-1}}{y^8}}$

$= \dfrac{3x^2}{4y^3} \cdot \dfrac{y^8}{x^{-1}}$

$= \dfrac{3x^{2+1}\,y^{8-3}}{4}$

$= \dfrac{3x^3 \cdot y^5}{4}$

Correct Answer : C

43. Solution:

Scientific notation form:

$A \times 10^n$ and $1 \le A < 10$

$= 0.00056712$

$= 5.6712 \times 10^{-4}$

Correct Answer : D

44. Solution:

Scientific notation form:

$A \times 10^n$ and $1 \leq A < 10$

$= 875$

$= 8.75 \times 10^2$

Correct Answer : A

45. Solution:

Scientific notation form:

$A \times 10^n$ and $1 \leq A < 10$

$= (4^3 \times 10^{-4})(2 \times 10^{-8})$

$= (64 \times 10^{-4})(2 \times 10^{-8})$

$= 128 \times 10^{-4-8}$

$= 128 \times 10^{-12}$

$= 1.28 \times 10^{-12+2}$

$= 1.28 \times 10^{-10}$

Correct Answer : B

46. Solution:

From figure Shaded portion: 8 and unshaded portion: 4

Ratio from shaded to unshaded is $\frac{8}{4} = 2$

Correct Answer : D

47. Solution:

If $x^{\frac{1}{4}} = 2$ then $x = 2^4$, $x = 16$

x must be 16

y must be 4

Correct Answer : D

48. Solution:

To find the LCM of 7 and 8 list the multiples.

Multiples of 7: 7, 14, 21, 28, 35, 42, 47, 56

Multiples of 8: 8, 16, 24, 32, 40, 48, 56

The lowest multiple that is common to 7 and 8 is 56. So the LCM of 7 and 8 is 56

Correct Answer : D

49. Solution:

$$\frac{\sqrt{x}}{2+\sqrt{x}} = \frac{\sqrt{x} \cdot (2-\sqrt{x})}{(2+\sqrt{x})(2-\sqrt{x})} = \frac{2\sqrt{x}-x}{4-x}$$

Correct Answer : A

50. Solution:

$$\frac{x+x+1+x+2+x+3+x+4}{5} = 28$$

$$\frac{5x+10}{5} = 28$$

$$x+2 = 28$$

$$x = 26$$

The greatest one $= x + 4$

$\qquad\qquad = 26 + 4$

$\qquad\qquad = 30$

Correct Answer : C

American Math Academy

ORDER OF OPERATIONS EXERCISES SOLUTIONS

1. Solution:

$6(0) + 25$

$= 0 + 25$

$= 25$

2. Solution:

$3(6 - 6) + 3^3$

$= 3 \cdot 0 + 3^3$

$= 0 + 3^3 = 32 = 3 \cdot 3 \cdot 3 = 27$

3. Solution:

$52 - 48 + 11$

$= 4 + 11$

$= 15$

4. Solution:

$6(21 - 6) - 2^3$

$= 6(15) - 8$

$= 90 - 8 = 82$

5. Solution:

$6(22 - 21) + 21 - 8$

$= 6(1) + 21 - 8$

$= 6 + 21 - 8 = 27 - 8 = 19$

6. Solution:

$25 + (-3) \times 16 \div 12$

$25 + (-48 \div 12)$

$= 25 - 4 = 21$

7. Solution:

$(125 \div 125 \times 3 - 15) + 1$

$= (1 \times 3 - 15) + 1 = (3 - 15) + 1 = -12 + 1$

$= -11$

8. Solution:

$(32 \div 8 \times 3 + 15) - 12$

$= (4 \times 3 + 15) - 12$

$= 12 + 15 - 12 = 15$

9. Solution:

$$\frac{3^4 \cdot 2^4}{3^1 \cdot 2^4} = \frac{3^4}{3^1} = 3^{4-1} = 27$$

10. Solution:

$$\frac{64 - (4)}{3(16 - 8) - 4} = \frac{60}{3(8) - 4} = \frac{60}{24 - 4} = \frac{60}{20} = 3$$

11. Solution:

$$\frac{32}{2^5} = \frac{32}{32} = 1$$

12. Solution:

$$\frac{5(4) + 9(10)}{4 + 2} = \frac{20 + 90}{6} = \frac{110}{6} = \frac{55}{3}$$

13. Solution:

$12 + 16 - 3(7 + 3)$

$= 12 + 16 - 3(10) = 12 + 16 - 30$

$= 28 - 30 = -2$

14. Solution:

$\frac{15}{5} - \sqrt{36} + 15$

$= 3 - 6 + 15$

$= -3 + 15$

$= 12$

15. Solution:

$(8 \div 8 \times 5 + 20) - 10$

$= (1 \times 5 + 20) - 10$

$= 25 - 10$

$= 15$

16. Solution:

$100 - 25 + \frac{6}{3}$

$= 100 - 25 + 2$

$75 + 2$

$= 77$

1. Solution:

$$\frac{-1 \times 5}{3 \times 5} + \frac{1 \times 3}{3 \times 3} = \frac{-5}{15} + \frac{3}{15} = \frac{-2}{15}$$

2. Solution:

$$\frac{1 \times 3}{4 \times 3} - \frac{1}{12} = \frac{3}{12} - \frac{1}{12} = \frac{2}{12} = \frac{1}{6}$$

3. Solution:

$$\frac{3 \times 7}{5 \times 7} - \frac{4 \times 5}{7 \times 5} = \frac{21}{35} - \frac{20}{35} = \frac{1}{35}$$

4. Solution:

$$\frac{7}{49} + \frac{6}{36} = \frac{1}{7} + \frac{1}{6} = \frac{1 \times 6}{7 \times 6} + \frac{1 \times 7}{6 \times 7}$$

$$= \frac{6}{42} + \frac{7}{42} = \frac{13}{42}$$

5. Solution:

$$\frac{5}{4} + \frac{7}{3} = \frac{5 \times 3}{4 \times 3} + \frac{7 \times 4}{3 \times 4} = \frac{15}{12} + \frac{28}{12} = \frac{43}{12}$$

6. Solution:

$$\frac{7}{2} - \frac{21}{5} = \frac{7 \times 5}{2 \times 5} - \frac{21 \times 2}{5 \times 2}$$

$$= \frac{35}{10} - \frac{42}{10} = \frac{-7}{10}$$

7. Solution:

$$\frac{\overset{2}{\cancel{16}} \times 1}{\underset{1}{\cancel{8}} \times 10} = \frac{2}{10} = \frac{1}{5}$$

8. Solution:

$$\frac{25}{8} \times \frac{1}{9} = \frac{25}{72}$$

9. Solution:

$$\frac{1}{18} \times \frac{4}{9} = \frac{4}{162} = \frac{2}{81}$$

10. Solution:

$$\frac{\overset{1}{11}}{2} \times \frac{3}{\underset{2}{22}} = \frac{3}{4}$$

11. Solution:

$$\frac{72 \div 18}{18 \div 18} = \frac{4}{1} = 4$$

12. Solution:

$$\frac{18}{\underset{1}{17}} = \frac{\overset{2}{34}}{1} = 18 \times 2 = 36$$

13. Solution:

$$\frac{-1}{12} \div -\frac{9}{4} = \frac{-1}{12} \times \left(-\frac{4}{9}\right) = \frac{4}{108} = \frac{2}{54} = \frac{1}{27}$$

14. Solution:

$$\overset{3}{33} \times \frac{3}{\underset{2}{22}} = \frac{9}{2}$$

15. Solution:

$$\frac{1}{12} + \frac{8 \times 12}{1 \times 12} = \frac{1}{12} + \frac{96}{12} = \frac{97}{12}$$

16. Solution:

$$\frac{1}{2} \times \frac{5}{3} = \frac{5}{6}$$

American Math Academy

1. Solution:

$4 - (-8) = 12$

2. Solution:

$-12 - 20 = -32$

3. Solution:

$16 \div (-6) = \dfrac{16}{-6} = \dfrac{-8}{3}$

4. Solution:

$-25 - 34 + 9$

$= -59 + 9 = -50$

5. Solution:

$-16 \div (24) = \dfrac{-16}{24} = \dfrac{-8}{12} = \dfrac{-4}{6} = \dfrac{-2}{3}$

6. Solution:

$15 - 14 + \dfrac{1}{2} = \dfrac{3}{2}$

7. Solution:

$\dfrac{5}{-15 + 10} = \dfrac{5}{-5} = -1$

8. Solution:

$\dfrac{-8 \div 8}{-(-7)} = \dfrac{-1}{7}$

9. Solution:

$15(-4) = -60$

10. Solution:

$\dfrac{2}{6} \times \dfrac{-3}{4} = \dfrac{-6}{24} = \dfrac{-1}{4}$

11. Solution:

$\dfrac{b}{a}$

12. Solution:

$\dfrac{-5}{3}$

13. Solution:

$\dfrac{-1}{4}\left(\dfrac{-5}{1}\right) = \dfrac{5}{4}$ reciprocal $= \dfrac{4}{5}$

14. Solution:

reciprocal of $\dfrac{-9}{5}$ is $\dfrac{-5}{9}$

15. Solution:

$-5(-8) = 40$

opposite of 40 is -40

16. Solution:

opposite of -24 is

24

1. Solution:

$2^{3+4+8} = 2^{15}$

2. Solution:

$3^{-4-12+16} = 3^{-16+16} = 3^0 = 1$

3. Solution:

$6^{-3-(-4)} = 6^{-3+4} = 6^1$

4. Solution:

$\dfrac{5^2 \times 5^3}{5^6} = \dfrac{5^5}{5^6} = \dfrac{1}{5^1}$

5. Solution:

$\left(\dfrac{x^3}{y^6}\right)^0 = 1$

6. Solution:

$(2^1 \, x^1 \, y^3)^3 = 2^3 \, x^3 \, y^9$

$= 8x^3 \, y^9$

7. Solution:

$3^{-2} \, x^{-2} \, y^{10} = \dfrac{y^{10}}{9x^2}$

8. Solution:

$\dfrac{x^{-6}}{3^{-2}} = \dfrac{9}{x^6}$

9. Solution:

$\dfrac{3^{4-5+12}}{3^{-7}} = \dfrac{3^{-1+12}}{3^{-7}} = \dfrac{3^{11}}{3^{-7}} = 3^{11-(-7)} = 3^{18}$

10. Solution:

$\dfrac{\overset{3}{\cancel{36}} x^2 y^3}{\cancel{12}x^4 y^2} = 3x^{2-4} \, y^{3-2}$

$= 3x^{-2} y^1 = \dfrac{3y}{x^2}$

11. Solution:

$\dfrac{\overset{1}{\cancel{3}}x^{-3} y^{-4}}{\underset{3}{\cancel{9}}x^{-5} y^{-6}} = \dfrac{1}{3} \cdot x^{-3+5} \, y^{-4+6} = \dfrac{1}{3}x^2 y^2$

12. Solution:

$\dfrac{x^1 y^3}{1} \cdot \dfrac{x^3 y^{-4}}{1} = x^4 \, y^{-1} = \dfrac{x^4}{y^1}$

13. Solution:

$y^{3-(-8)} \cdot y^{-5} = y^{11} \cdot y^{-5} = y^6$

14. Solution:

$2^{x+x+x} = 2^{3x}$

15. Solution:

$3^x + 3^x + 3^x$

$3^x (1 + 1 + 1)$

$= 3^x \cdot 3^1$

3^{x+1}

16. Solution:

$\dfrac{2^{2x}}{(2^4)^{3x}} \times 2^{5x}$

$= \dfrac{2^{2x}}{2^{12x}} \cdot 2^{5x} = \dfrac{1}{2^{10x}} \times 2^{5x}$

$= 2^{5x-10x} = 2^{-5x} = \dfrac{1}{2^{5x}}$

ABSOLUTE VALUE AND INEQUALITY
EXERCISES SOLUTIONS

1. Solution:

$|3x - 6| = 9$

$3x - 6 = 9$ or $3x - 6 = -9$

$3x = 15 \qquad\qquad 3x = -3$

$x = 5 \qquad\qquad\quad x = -1$

2. Solution:

No answer is possible

3. Solution:

$4|8 - 7| = 4 \times 1 = 4$

4. Solution:

$\dfrac{1}{11}|27 - 16| = \dfrac{1}{11}|11| = \dfrac{11}{11} = 1$

5. Solution:

$\dfrac{|8 + 4|}{2^4} = \dfrac{12}{16} = \dfrac{3}{4}$

6. Solution:

$3x + 5 > 9, \qquad\qquad 3x > 4$

$x > \dfrac{4}{3}$

7. Solution:

$2x - 10 < 12$ or $-12 < 2x - 10$

$2x < 22 \qquad\qquad -2 < 2x$

$x < 11 \qquad\qquad -1 < x$

8. Solution:

$3x - 6 > 18$ or $-18 > 3x - 6$

$3x > 24 \qquad\qquad -12 > 3x$

$x > 8 \qquad\qquad -4 > x$

9. Solution:

$$\frac{-3-2x}{3} < \frac{x-4}{1} = 3 - 2x < 3x - 12$$

$$3 + 12 < 3x + 2x$$

$$15 < 5x$$

$$3 < x$$

10. Solution:

$$3x - 1 \geq 2x - 10$$

$$3x - 2x > 1 - 10$$

$$x > -9$$

11. Solution:

$$2x + 6 < -20$$

$$2x < -20 - 6$$

$$2x < -26$$

$$x < -13$$

12. Solution:

$$3(x + 6) < 5 \cdot 4x$$

$$3x + 18 < 20x$$

$$18 < 17x$$

$$\frac{18}{17} < x$$

13. Solution:

$$6(2x - 7) \leq 4(3x + 5)$$

$$12x - 42 \leq 12x + 20$$

$$-42 \leq 20 \text{ (Many Solutions)}$$

14. Solution:

$$4|x - 8| > 24$$

$$|x - 8| > 6$$

$$x - 8 > 6 \quad \text{or} \quad x - 8 < -6$$

$$x > 14 \qquad\qquad x < 2$$

15. Solution:

$$3x - 5 = 5x \quad \text{or} \quad 3x - 5 = -5x$$

$$-5 = 5x - 3x \qquad -5 = -8x$$

$$-5 = 2x \qquad\qquad \frac{5}{8} = x$$

$$\frac{-5}{2} = x$$

16. Solution:

$$\frac{1}{3}|x| = 13$$

$$|x| = 39$$

$$x = \mp 39$$

LAWS OF RADICALS
EXERCISES SOLUTIONS

1. Solution:

$$\sqrt{12^2} = 12$$

2. Solution:

$$\sqrt{4^2} + \sqrt{5^2} - \sqrt{6^2}$$

$$= 4 + 5 - 6 = 9 - 6 = 3$$

3. Solution:

$$\sqrt{400 \times 2} + \sqrt{2 \times 100}$$

$$= 20\sqrt{2} + 10\sqrt{2} = 30\sqrt{2}$$

4. Solution:

$$\sqrt[3]{\frac{2^3}{3^3}} = \sqrt[3]{\left(\frac{2}{3}\right)^3} = \frac{2}{3}$$

5. Solution:

$$\frac{\sqrt{5^2}}{\sqrt[3]{3^3}} = \frac{5}{3}$$

6. Solution:

$$\sqrt{9 \times 2} - \sqrt{36 \times 2}$$

$$= 3\sqrt{2} - 6\sqrt{2} = -3\sqrt{2}$$

7. Solution:

$$\sqrt[3]{8 \times 10^6} = \sqrt[3]{2^3 \times 10^6} = 2 \times 10^2$$

$$= 200$$

8. Solution:

$$\frac{6\sqrt{5} \times \sqrt{3}}{\sqrt{3}} \times \frac{\sqrt{3}}{\sqrt{3}} = \frac{6\sqrt{15}}{\sqrt{9}} = \frac{6\sqrt{15}}{3}$$

$$= 2\sqrt{15}$$

9. Solution:

$$\sqrt{a \cdot a \cdot a} = \sqrt{a^3} = a\sqrt{a}$$

10. Solution:

$$\frac{1}{\sqrt{x \cdot x \cdot x \cdot x}} = \frac{1}{\sqrt{x^4}} = \frac{1}{x^2}$$

11. Solution:

$$\sqrt{11^2} + \sqrt{8^2} - \sqrt{12^2}$$

$$= 11 + 8 - 12 = 19 - 12 = 7$$

12. Solution:

$$3\sqrt{3} - 2\sqrt{3} + \sqrt{3}$$

$$= \sqrt{3} + \sqrt{3} = 2\sqrt{3}$$

13. Solution:

$$\frac{2\sqrt{a}}{\sqrt{a}} \cdot \sqrt{a} = 2\sqrt{a}$$

14. Solution:

$$2\sqrt{4 \cdot 5} - \sqrt{5}$$

$$= 2 \cdot 2\sqrt{5} - \sqrt{5} = 4\sqrt{5} - \sqrt{5} = 3\sqrt{5}$$

15. Solution:

$$3\sqrt{9 \cdot 5} - 9\sqrt{5} + \sqrt{25 \cdot 3}$$

$$= 3 \cdot 3\sqrt{5} - 9\sqrt{5} + 5\sqrt{3}$$

$$= 9\sqrt{5} - 9\sqrt{5} + 5\sqrt{3} = 5\sqrt{3}$$

16. Solution:

$$3\sqrt{3} + \sqrt{9 \cdot 3} + \sqrt{4 \cdot 3}$$

$$3\sqrt{3} + 3\sqrt{3} + 2\sqrt{3} = 8\sqrt{3}$$

American Math Academy

COORDINATE PLANE
EXERCISES SOLUTIONS

Name the coordinates for each given point. Give the quadrant for each point.

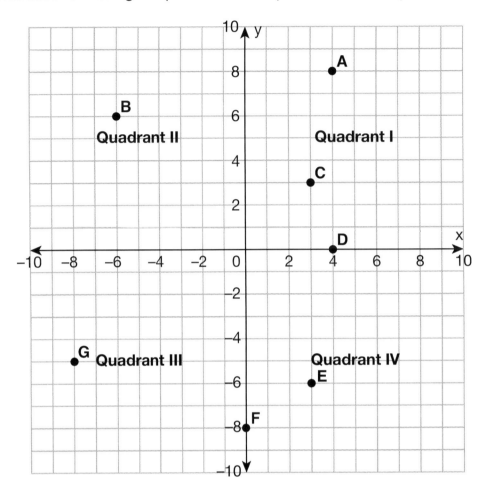

Coordinates	Point	Quadrant
(4, 8)	A	I
(–6, 6)	B	II
(3, 3)	C	I
(4, 0)	D	No quadrant
(3, –6)	E	IV
(0, –8)	F	No quadrant
(–8, –5)	G	III

FACTORS & MULTIPLES
(GCF AND LCM) EXERCISES SOLUTIONS

Circle the numbers from list that are prime

1	(2)	(3)	4	(5)
6	(7)	8	9	10
(11)	12	(13)	14	15
16	(17)	18	(19)	20
21	22	(23)	24	25

Circle the numbers that are composite

1	2	3	(4)	5
(6)	7	(8)	(9)	(10)
11	(12)	13	(14)	(15)
(16)	17	(18)	19	(20)
(21)	(22)	23	(24)	(25)

Find the greatest common factor (GCF) and least common multiple (LCM) of each pair of **numbers**

	GCF		LCM	
15, 45	15: 1, 3, 5, 15 45: 1, 3, 5, 15, 45	(15)	15: 15, 30, 45 45: 45, 90	(45)
16, 24	16: 1, 2, 4, 8, 16 24: 1, 2, 3, 4, 6, 8, 12, 24	(8)	16: 16, 32, 48 24: 24, 48	(48)
18, 48	18: 1, 2, 3, 6, 9, 18 48: 1, 2, 3, 4, 6, 8, 12, 24, 48	(6)	18: 18, 36, 54, 72, 90, 108,126, 144 48: 48, 96, 144	(144)
5, 45	5: 1, 5 45: 1, 3, 5, 9, 45	(5)	5: 5, 10, 15, 20, 25, 30, 35, 40, 45 45: 45	(45)
11, 88	11: 1, 11 88: 1, 8, 11	(11)	11: 11, 22, 33, 44, 55, 66, 77, 88 88: 88	(88)
5, 10, 15	5: 1, 5 10: 1, 5, 15 15: 1, 3, 5, 15	(5)	5: 5, 10, 15, 20, 25, 30 10: 10, 20, 30 15: 15, 30	(30)
16, 24, 36	16: 1, 2, 4, 8, 16 24: 1, 2, 3, 4, 6, 8, 12, 24 36: 1, 2, 3, 4, 6, 9, 12, 18, 36	(4)	16: 16, 32, ... 144 24: 24, 48, ... 144 36: 36, 72, ... 144	(144)

SCIENTIFIC NOTATION
EXERCISES SOLUTIONS

1. **Solution:**

   $125 \times 10^5 = 1.25 \times 10^2 \times 10^5$

   $= 1.25 \times 10^7$

2. **Solution:**

   $1.23 \times 10^{-2} \times 10^5$

   1.23×10^3

3. **Solution:**

   $9.8 \times 10^{-1} \times 10^{-6} = 9.8 \times 10^{-7}$

4. **Solution:**

   $1.45 \times 10^5 \times 10^3 = 1.45 \times 10^8$

5. **Solution:**

   $4.57 \times 10^{-2} \times 10^{-9}$

   $= 4.57 \times 10^{-11}$

6. **Solution:**

   $8.69 \times 10^1 \times 10^{-12}$

   $= 8.69 \times 10^{-11}$

7. **Solution:**

   $967 (10,000) = 9670,000$

8. **Solution:**

   $= 0.00457$

9. **Solution:**

   $21,900$

10. **Solution:**

    $1.5 \times 1 = 1.5$

11. **Solution:**

    $0.04 (1000,000) = 40000$

12. **Solution:**

    $3 \cdot 45 (10,000,000) = 34,500,000$

13. **Solution:**

    1.8×10^0

14. **Solution:**

    $4 \cdot 44 \times 10^{-2}$

15. **Solution:**

    $3 \cdot 784 \times 10^0$

16. **Solution:**

    $8 \cdot 9 \times 10^7$

17. **Solution:**

    $3 \cdot 67 \times 10^0$

18. **Solution:**

    $2 \cdot 34 \times 10^5$

1. Solution:

   $$\frac{2\cancel{0}}{3\cancel{0}} = \frac{2}{3}$$

2. Solution:

   $$\frac{8}{35}$$

3. Solution:

   $$\frac{17}{68} = \frac{1}{4}$$

4. Solution:

   $$\frac{72}{66} = \frac{36}{33} = \frac{12}{11} = 1\frac{1}{11}$$

5. Solution:

   $4 \neq 12$ No

6. Solution:

   $2 \cdot 12 = 24 \cdot 1$

   $24 = 24$ Yes

7. Solution:

   $2 \cdot 54 = 6 \cdot 18$

   $108 = 108$ Yes

8. Solution:

   $7 \cdot 120 = 90 \cdot 35$

   $840 \neq 3150$ No

9. Solution:

   $54x = 36$

   $$x = \frac{36}{54} = \frac{4}{6} = \frac{2}{3}$$

10. Solution:

    $7x = 3 \cdot 54$

    $7x = 162 \qquad x = \dfrac{162}{7}$

11. Solution:

    $-25x = 75$

    $$x = \frac{75}{-25} = -3$$

12. Solution:

    $4(x + 5) = 8 \cdot 9$

    $4x + 20 = 72$

    $4x = 52 \qquad x = 13$

13. Solution:

    $$\frac{x}{12} = \frac{1}{4} \cdot \frac{5}{3}$$

    $\dfrac{x}{\cancel{12}} = \dfrac{5}{\cancel{12}} \qquad x = 5$

14. Solution:

    $$\frac{1}{\underset{2}{8}} \quad \frac{\overset{3}{\cancel{12}}}{16}$$

    $$x = \frac{1}{2} \times \frac{3}{16} = \frac{3}{32}$$

15. Solution:

    $3x + 5x = 32 \qquad 8x = 32 \qquad x = 4$

    Boys: $5x = 5 \cdot 4 = 20$

16. Solution:

    $y = kx \qquad 20 = 5 \cdot k$

    $\qquad\qquad\qquad 4 = k$

    $y = kx \qquad y = 4x$

    $y = 4 \cdot 3 \qquad y = 12$

17. Solution:

    $y = \dfrac{k}{x} \qquad 60 = \dfrac{k}{12} \qquad k = 720$

    $$y = \frac{720}{18} = 40$$

1. Solution:

$D = r \cdot t$

$30 = 1 \cdot r$

$30 = r$

2. Solution:

$D = r \cdot t$

$20 = 5 \cdot r$

$4 = r$

3. Solution:

$35 = r \cdot 7$

$5 = r$

4. Solution:

$10 = \dfrac{25}{100} \cdot x$

$10 = \dfrac{1}{4} \cdot x$

$40 = x$

5. Solution:

$\dfrac{45}{10\cancel{0}} \cdot 4\cancel{0} = x$

$18 = x$

6. Solution:

$\dfrac{x}{10\cancel{0}} \cdot 3\cancel{0} = 6$

$3x = 60$

$x = 20$

7. Solution:

$x = \dfrac{10}{10} \cdot 90$

$x = \dfrac{900}{100} = 9$

8. Solution:

$\dfrac{\$480}{24} = \$20$

9. Solution:

Original price = 100x after 20% discount

$80x = 45$

$x = \dfrac{45}{80}$

Original price $= 100x = \dfrac{100 \times 45}{80}$

$= \$56.25$

10. Solution:

Discount price $= \dfrac{40 \times 25}{100} = \dfrac{1000}{100} = \$10$

Sale price: $\$40 - \$10 = \$30$

1. $x = \dfrac{4^3}{2^7} = \dfrac{2^6}{2^7} = 2^{-1} = \dfrac{1}{2}.$

   Correct Answer : C

2. $\sqrt[3]{64} = \sqrt[3]{2^6} = 2^{\frac{6}{3}} = 2^2 = 4.$

   Correct Answer : B

3. $\sqrt{25} - \sqrt{36} + \sqrt{100} = 5 - 6 + 10 = 9$

   Correct Answer : D

4. $\sqrt{(16)^2} + \sqrt{2^2} - (-2)^3 = 16 + 2 + 8 = 26$

   Correct Answer : D

5. $x\sqrt{25} = 45$

   $5x = 45$

   $x = 9$

   Correct Answer : B

6. $\dfrac{\sqrt{250}}{\sqrt{10}} \cdot \sqrt{49} = \sqrt{\dfrac{250}{10}} \cdot \sqrt{49} = \sqrt{25} \cdot \sqrt{49} = 5 \cdot 7 = 35$

   Correct Answer : C

7. $x = 1 + \sqrt{2}$

   $y = 1 - \sqrt{2}$

   $x \cdot y = (1 + \sqrt{2}) \cdot (1 - \sqrt{2}) = 1 - 2 = -1$

   Correct Answer : B

8. $\dfrac{8^2 \cdot 16^3}{2^{10}} = \dfrac{2^6 \cdot 2^{12}}{2^{10}} = \dfrac{2^{18}}{2^{10}} = 2^8$

   Correct Answer : C

9. $x^{\frac{2}{7}} = \sqrt[7]{x^2}$

   Correct Answer : C

10. If $x^5 = 32$, then

    $x^5 = 2^5$

    $x = 2$, then $x^2 = 4$

    Correct Answer : C

MIXED REVIEW TEST I SOLUTIONS

11. Melisa spent $1\frac{3}{8}=\frac{11}{8}$ hours

John spent $=\frac{11}{8}\cdot\frac{1}{4}=\frac{11}{32}$ hours.

Correct Answer : D

12. Bank checking account:
$258 + $140 − $128 = $270.

Correct Answer : A

13. Sum of the three consecutive integers:
x + x + 1 + x + 2
x + x + 1 + x + 2 = 96
$$3x + 3 = 96$$
$$3x = 93$$
$$x = 31$$
The lowest one is 31.

Correct Answer : D

14. $4^{x-4} = 8^{x-6}$
$2^{2x-8} = 2^{3x-18}$
$2x - 8 = 3x - 18$
$x = 18 - 8$
$x = 10$

Correct Answer : B

15. $\frac{2x-7}{5}=\frac{2x-9}{3}$, then cross–multiply.
$3(2x - 7) = 5(2x - 9)$
$6x - 21 = 10x - 45$
$45 - 21 = 4x$
$24 = 4x$
$6 = x$

Correct Answer : D

16. If $x = \dfrac{2^3}{\sqrt{64}} = \dfrac{8}{8} = 1$

Correct Answer : A

17. If y − 2x = 10, then
$y = 2x + 10$
$x = \dfrac{y-10}{2}$
$4x = 4\left(\dfrac{y-10}{2}\right) = 2(y-10) = 2y - 20$

Correct Answer : C

18. If x > 0 and $(3x − 5)^2 = 49$
$3x - 5 = 7$
$3x = 12$
$x = 4$

Correct Answer : B

American Math Academy

AMERICAN MATH
ACADEMY

19. $x + 2y = 3$

$2^x \cdot 4^y = 2^x \cdot 2^{2y} = 2^{x+2y} = 2^3 = 8$

Correct Answer : D

20. 24 pens of $\dfrac{3}{4} = 24 \cdot \dfrac{3}{4} = 18$

Correct Answer : C

21. If $20 = 5\left(\dfrac{x}{3} - 4\right)$, then

$4 = \dfrac{x}{3} - 4$

$8 = \dfrac{x}{3}$

$24 = x$

Correct Answer : A

22. Since a and b are positive integers and

$\sqrt{a} = b^2 = 4$.

a can be 16

b can be 2

Then, $a - b = 16 - 2 = 14$

Correct Answer : D

23. $F = \dfrac{9}{5}C + 32$, since $C = 60$

$F = \dfrac{9}{5}(60) + 32$

$F = 9 \cdot 12 + 32$

$F = 108 + 32$

$F = 140F.$

Correct Answer : D

American Math Academy

MIXED REVIEW TEST II SOLUTIONS

1. $a = 3b + 6 \rightarrow a - 3b = 6$

$3a - 9b = 18 \rightarrow a - 3b = 6$

$a - 3b = a - 3b$

$0 = 0$, then the systems have many/infinity solutions.

Correct Answer : D

2. $|2x - 5| = 7$ Since $x > 0$, $2x - 5 = 7$

$2x = 7 + 5$

$2x = 12$

$x = 6$

Correct Answer : C

3. Rule: $a^2 - b^2 = (a - b)(a + b)$

$81x^2 - 4y^2 = (9x - 2y)(9x + 2y)$

Correct Answer : A

4. If $5a - 3b = -6$ and $a = 6$, then $5a - 3(6) = -6$

$5a - 18 = -6$

$5a = 18 - 6$

$5a = 12$

$a = 2.4$

Correct Answer : 2.4

5. If $x^3 = 27$, then $x^3 = 3^3 \rightarrow x = 3$

$x^2 = 3^2 = 9$

Correct Answer : B

6. If the ratio of $\frac{1}{3} : \frac{1}{b}$ is equal to $\frac{1}{4} : \frac{1}{2}$,

then $\dfrac{\frac{1}{3}}{\frac{1}{b}} = \dfrac{\frac{1}{4}}{\frac{1}{2}}$.

$\frac{b}{3} = \frac{1}{2}$

$2b = 3$

$b = \frac{3}{2}$

Correct Answer : B

7. If $3^{2x-4} = 27^{x-6}$, then $3^{2x-4} = 3^{3x-18}$. Since the bases are equal, the powers need to be equal as well.

$2x - 4 = 3x - 18$

$-4 + 18 = 3x - 2x$

$14 = x$

Correct Answer : C

110

AMERICAN MATH ACADEMY

American Math Academy

8. If $\dfrac{3x-1}{2x+1} = \dfrac{5}{3} \rightarrow$ cross multiply

$3(3x - 1) = 5(2x + 1)$

$9x - 3 = 10x + 5$

$-3 - 5 = 10x - 9x$

$-8 = x$

Correct Answer : D

9. If $x = \dfrac{3^3}{\sqrt{81}}$, then $x = \dfrac{3^3}{9} = \dfrac{27}{9} = 3$

Correct Answer : C

10. $x + ay - 6 = 0$

Slope of equation $\rightarrow \dfrac{-1}{a} = \dfrac{1}{3}$

$a = -3$

Correct Answer : C

11. Slope of points of line

$L = \dfrac{y_2 - y_1}{x_2 - x_1} = \dfrac{8 - 5}{2 - (-1)} = \dfrac{3}{3} = 1$

Correct Answer : A

12. $x - 2y = 12 \rightarrow -2(x - 2y = 12) = -2x + 4y = -24$

$\begin{array}{r} -2x + 4y = -24 \\ +\underline{3x - 4y = 18} \\ x = -6 \end{array}$

Correct Answer : C

13. $\dfrac{2x-1}{3} - \dfrac{x}{2} = \dfrac{3}{2}$

$\dfrac{2(2x-1)}{6} - \dfrac{3x}{6} = \dfrac{9}{6}$

$2(2x - 1) - 3x = 9$

$4x - 2 - 3x = 9$

$x - 2 = 9$

$x = 11$

Correct Answer : B

14. The ratio of $\dfrac{1}{3}$ to $\dfrac{4}{3} = \dfrac{\frac{1}{3}}{\frac{4}{3}} = \dfrac{3}{12} = \dfrac{1}{4}$

Correct Answer : D

American Math Academy

15. $4^{15} \cdot 4^2 = 4^{-n} \cdot 4^m \cdot 4^7$

$4^{15+2} = 4^{-n+m+7}$

$17 = -n + m + 7$

$17 - 7 = m - n$

$10 = m - n$

Correct Answer : A

16. $x^{\frac{3}{4}} = \sqrt[4]{x^3}$

Correct Answer : C

17. $\sqrt{3} \cdot \sqrt{16} = \sqrt{3} \cdot 4 = 4\sqrt{3}$

Correct Answer : A

18. $\sqrt[3]{0.027} = \sqrt[3]{\dfrac{27}{1000}} = \sqrt[3]{\dfrac{3^3}{10^3}} = \dfrac{3}{10} = 0.3$

Correct Answer : B

American Math Academy

1. Solution:

$5x = 25$

$x = 5$

2. Solution:

$3x = 45$

$x = 15$

3. Solution:

$4x = -1$

$x = \dfrac{-1}{4}$

4. Solution:

$12x = 24$

$x = 2$

5. Solution:

$-6x = -30$

$6x = 30$

$x = 5$

6. Solution:

$7x = 28$

$x = 4$

7. Solution:

$\dfrac{x}{3} = 15$

$x = 3 \cdot 15$

$x = 45$

8. Solution:

$\dfrac{-3x}{8} = 6$

$-3x = 48$

$x = -16$

9. Solution:

$\dfrac{3x \cdot 2}{7 \cdot 2} - \dfrac{1}{14} = 9$

$\dfrac{6x}{14} - \dfrac{1}{14} = 9$

$6x - 1 = 9 \cdot 14$

$6x = 127$

$x = \dfrac{127}{6}$

10. Solution:

$\dfrac{x}{-7} = 7 + 9$

$-\dfrac{x}{7} = 16$

$x = -112$

11. Solution:

$$3x + 12 = -9$$

$$3x = -21$$

$$x = -7$$

12. Solution:

$$8x = 40$$

$$x = 5$$

13. Solution:

$$\frac{x \times 4}{3 \times 4} - \frac{1 \times 3}{4 \times 3} = \frac{1}{2}$$

$$\frac{4x}{12} - \frac{3}{12} = \frac{1}{12}$$

$$4x - 3 = 1$$

$$4x = 4$$

$$x = 1$$

14. Solution:

$$\frac{x}{7} + 6 = -1$$

$$\frac{x}{7} = -7$$

$$x = -49$$

15. Solution:

$$\frac{-1 \times 2}{3 \times 2} - \frac{x \times 3}{2 \times 3} = 7$$

$$\frac{-2 - 3x}{6} = 7$$

$$2 - 3x = 42$$

$$-3x = 44$$

$$x = \frac{-44}{3}$$

16. Solution:

$$4x - 3 = 9$$

$$4x = 12$$

$$x = 3$$

17. Solution:

$$\frac{x}{12} - \frac{x}{6} = 5$$

$$\frac{x - 2x}{12} = 5$$

$$\frac{-x}{12} = 5$$

$$x = -5 \cdot 12$$

$$x = -60$$

18. Solution:

$$3 + \frac{x}{7} = -2$$

$$\frac{x}{7} = -5$$

$$x = -35$$

American Math Academy

AMERICAN MATH
ACADEMY

1. Solution:

$2x - x = 10 + 5$

$x = 15$

2. Solution:

$5x - 4x = 15 + 12$

$x = 27$

3. Solution:

$4x = -3x = 5 - 6$

$x = -1$

4. Solution:

$-3x + 5x = -24 - 12$

$2x = -36$

$x = -18$

5. Solution:

$-6x - 10x = -10 - 20$

$-16x = -30$

$x = \dfrac{30}{16} = \dfrac{15}{8}$

6. Solution:

$-28 = 2x$

$x = -14$

7. Solution:

$\dfrac{x}{3} - \dfrac{x}{4} = 2 + 5$

$\dfrac{4x - 3x}{12} = 7$

$x = 7 \cdot 12$

$x = 84$

8. Solution:

$\dfrac{-1}{2} + \dfrac{1}{2} = 3x + 2x$

$0 = 5x$

$x = 0$

9. Solution:

$\dfrac{3}{7} - \dfrac{9}{7} = \dfrac{2x}{7} + \dfrac{x}{7}$

$\dfrac{-6}{7} = \dfrac{3x}{7}$

$-6 = 3x$

$-2 = x$

10. Solution:

$-x - 7 = -x + 5$

$-7 \neq 5$, No solution

11. Solution:

$$3x + 2x = -15$$

$$5x = -15$$

$$x = -3$$

12. Solution:

$$2x + 20 = 5x - 20$$

$$40 = 3x$$

$$x = \frac{40}{3}$$

13. Solution:

$$\frac{x-2}{10} = \frac{x}{5}$$

$$5x - 10 = 10x$$

$$-10 = 5x$$

$$-2 = x$$

14. Solution:

$$2x + 6 = -x + 8$$

$$3x = 2$$

$$x = \frac{2}{3}$$

15. Solution:

$$\frac{-12-2x}{2} = 2x$$

$$-12 - 2x = 4x$$

$$-12 = 6x$$

$$-2 = x$$

16. Solution:

$$4x - 3 = 2x - 3$$

$$2x - 3 = -3$$

$$2x = 0$$

$$x = 0$$

17. Solution:

$$\frac{x}{18} - \frac{x}{9} = \frac{x}{2}$$

$$\frac{x}{18} - \frac{x}{9} - \frac{x}{2} = 0$$

$$\frac{x - 2x - 9x}{18} = 0$$

$$\frac{-10x}{18} = 0$$

$$x = 0$$

18. Solution:

$$3 + 5 = x - \frac{x}{3}$$

$$8 = \frac{2x}{3}$$

$$24 = 2x$$

$$x = 12$$

1. Solution:

$3(8) + 3(5)$

$= 24 + 15 = 39$

2. Solution:

$5(7) + 5(2) = 35 + 10$

$= 45$

3. Solution:

$ab + ac$

4. Solution:

$4(x) + 4(3)$

$= 4x + 12$

5. Solution:

$-3(3x) - 3(-5)$

$= -9x + 15$

6. Solution:

$-2(-x) + (-2)(-5)$

$= 2x + 10$

7. Solution:

$-2(-x) - 2(-4)$

$= 2x + 8$

8. Solution:

$4(x) + 4(7)$

$= 4x + 28$

9. Solution:

$\frac{1}{2}(x) - \frac{1}{2}(12)$

$\frac{1}{2}(x) - 6$

10. Solution:

$\frac{x}{3}(x) + \frac{x}{3}(-6) = \frac{x^2}{3} - 2x$

American Math Academy

11. Solution:

$3x (a) + 3x(2b)$

$= 3ax + 6bx$

12. Solution:

$\dfrac{1}{3}(6x) + \dfrac{1}{3}(-9y)$

$= 2x - 3y$

13. Solution:

$-3(x) - 3(-2y) - 3(4z)$

$= -3x + 6y - 12z$

14. Solution:

$3x - 6 - 2x + 8$

$= 3x - 2x - 6 + 8 = x + 2$

15. Solution:

$5x(2x) + 5x(1) - 1(2x) - 1(1)$

$= 10x^2 + 5x - 2x - 1$

$= 10x^2 + 3x - 1$

16. Solution:

$a(a) + a(b) - b(a) - b(b)$

$= a^2 + \cancel{ab} - \cancel{ab} - b^2 = a^2 - b^2$

17. Solution:

$x(x) - x(y) + y(x) - y(y)$

$= x^2 - \cancel{xy} + \cancel{xy} - y^2 = x^2 - y^2$

18. Solution:

$a^2(a^2) - a^2(b^2) - b^2(a^2) - b^2(b^2)$

$= a^4 + \cancel{a^2b^2} - \cancel{a^2b^2} - b^4 = a^4 - b^4$

19. Solution:

$\dfrac{1}{x}\left(\dfrac{1}{x}\right) - \dfrac{1}{x}(x) + x\left(\dfrac{1}{x}\right) - x(x)$

$= \dfrac{1}{x^2} - \cancel{1} + \cancel{1} - x^2 = \dfrac{1}{x^2} - x^2$

20. Solution:

$a^2(a^2) - a^2(b^2)$

$= a^4 - a^2b^2$

American Math Academy

1. Solution:

$y = mx + b$

$2x - 3y = 12$

$2x - 12 = 3y$

$\dfrac{2}{3}x - 4 = y$

$m_1 = \dfrac{2}{3}$

$m_2 = \dfrac{3}{12} = \dfrac{1}{4}$ Neither.

2. Solution:

$y = \dfrac{-2}{3}x + 6$

$m_1 = \dfrac{-2}{3}$

$3x - 18 = 2y$

$\dfrac{3}{2}x - 9 = y$

$m_2 = \dfrac{3}{2}$

since $m_1 \cdot m_2 = -1$

lines are perpendicular.

3. Solution:

$y = mx + b$

$3x - 12 = 4y$

$\dfrac{3x}{4} - 3 = y$

$m_1 = \dfrac{3}{4}$

$m_2 = \dfrac{-4}{3}$

Since $m_1 \cdot m_2 = -1$, Perpendicular.

4. Solution:

$y = m + +b$

$y = -x + 6 \longrightarrow m_1 = -1$

$y = x + 0 \longrightarrow m_2 = 1$

So, lines are perpendicular.

6. Solution:

$y = mx + b$

$3x - 6 = 6y, \ \dfrac{3}{6}x - \dfrac{6}{6} = y$

$m_1 = \dfrac{3}{6} = \dfrac{1}{2}$

$y = -2x$

$m_2 = -2$

Since the product of the slopes is equal to -1, then lines are perpendicular.

5. Solution:

$y = mx + b$

$m_1 = -2, \ m_2 = -2$

since slopes are equal, lines are parallel.

SOLVING EQUATIONS INVOLVING PARALLEL AND PERPENDICULAR LINES EXERCISES SOLUTIONS

7. Solution:

$m_1 = 4$ and $m_1 /\!/ m_2$, so $m_2 = 4$

$y = mx + b$

$y = 4x + b$, $3 = 4(2) + b$, $b = -5$

$y = 4x - 5$

9. Solution:

$m_1 = 5$ and $m_1 /\!/ m_2$, so $m_2 = 5$

$y = 5x + b$

$-4 = 5(-1) + b$

$-4 = -5 + b$

$b = 1$

$y = 5x + 1$

11. Solution:

$m_1 = -1$

Since $m_1 \cdot m_2 = -1$, then $m_2 = 1$

$y = x + b$

$-3 = 2 + b$

$b = -5$

$y = x - 5$

13. Solution:

$y = x + a$

$m_1 = -1$

since $m_1 \cdot m_2 = -1$, $m_2 = 1$

$y = x + b$

$8 = 4 + b$

$4 = b$

$y = x + 4$

8. Solution:

$m = \dfrac{1}{2}$ and $m_1 /\!/ m_2$, so $m_2 = \dfrac{1}{2}$

$y = \dfrac{1}{2}x + b$, $\quad 5 = \dfrac{1}{2}(1) + b$

$10 = 1 + 2b$, $2b = 9$

$b = \dfrac{9}{2}$

$y = \dfrac{1}{2}x + \dfrac{9}{2}$

10. Solution:

$m_1 = \dfrac{2}{3}$ and $m_1 /\!/ m_2$, so $m_2 = \dfrac{2}{3}$

$y = \dfrac{2}{3}x + b$, $3 = \dfrac{2}{3}(1) + b \qquad b = \dfrac{7}{3}$

$y = \dfrac{2}{3}x + \dfrac{7}{3}$

12. Solution:

$m_1 = \dfrac{3}{2}$,

Since $m_1 \cdot m_2 = -1$

$m_2 = \dfrac{-2}{3}$

$y = \dfrac{-2}{3}x + b$, $\qquad -4 = \dfrac{-2}{3}(-1) + b$,

$b = \dfrac{-14}{3}$ $\qquad y = \dfrac{-2}{3}x - \dfrac{14}{3}$

14. Solution:

$y = \dfrac{1}{2}x - \dfrac{5}{2} \qquad m_1 = \dfrac{1}{2}$

Since $m_1 \cdot m_2 = -1$, then $m_2 = -2$

$y = -2x + b$, $\quad -2 = -2(-1) + b$

$-2 = 2 + b$, $b = -4$

$y = -2x - 4$

American Math Academy

AMERICAN MATH
ACADEMY

1. Solution:

$4 \ (2x + y = 20)$

$3x - 4y = 19$

$8x + 4\cancel{y} = 80$

$+ \ \dfrac{3x - 4\cancel{y} = 19}{}$

$11x = 99, \ x = 9$

$2 \cdot 9 + y = 20$

$18 + y = 20, \ y = 2$

2. Solution:

$x + 2\cancel{y} = 15$

$+ \ \dfrac{3x - 2\cancel{y} = 17}{}$

$4x = 32, \qquad x = 8$

$x + 2y = 15, \qquad 8 + 2y = 15$

$2y = 7, \qquad y = \dfrac{7}{2}$

3. Solution:

$3x + 2y = 10$

$-2 \ (x + y = 14)$

$3x + 2y = 10$

$+ \ \dfrac{-2x - 2y = -28}{}$

$x = -18$

$x + y = 14, \ -18 + y = 14$

$y = 32$

4. Solution:

$2x + 6y = 18$

$6 \ (2x - y = 4)$

$2x + 6\cancel{y} = 18$

$+ \ \dfrac{12x - 6\cancel{y} = 24}{}$

$14x = 42, \qquad x = 3$

$2x - y = 4, \qquad 2 \cdot 3 - y = 4$

$6 - y = 4, \qquad y = 2$

5. Solution:

$x - \dfrac{1}{2}y = 16$

$+ \ \dfrac{x + \dfrac{1}{2}y = 6}{}$

$2x = 22 \qquad x = 11$

$11 + \dfrac{1}{2}y = 6, \qquad \dfrac{1}{2}y = -5, \qquad y = -10$

6. Solution:

$7x - y = 118$

$- \ (2x - y = 8)$

$7x - y = 118$

$+ \ \dfrac{-2x + y = -8}{}$

$5x = 110, \qquad x = 22$

$2x - y = 8, \qquad 2 \cdot 22 - y = 8$

$y = 36$

American Math Academy

7. Solution:

$2y + 2 - 4y = 8$

$-2y + 2 = 8,$ $-2y = 6,$

$y = -3$

$x - 4y = 8,$ $x - 4(-3) = 8,$

$x = -4$

8. Solution:

$x + 2(x + 6) = 18$ $y = x + 6$

$x + 2x + 12 = 18$ $y = 2 + 6$

$3x + 12 = 18$ $y = 8$

$3x = 6$

$x = 2$

9. Solution:

$3y + y = 20,$ $4y = 20$

$y = 5$

$x = 3y,$ $x = 3 \cdot 5 = 15$

10. Solution:

$2x - 3y = 0$ $y = 12 - 2x$

$2x - 3(12 - 2x) = 0$ $2x - 36 + 6x = 0$

$8x = 36$

$x = \dfrac{36}{8} = \dfrac{9}{2}$

$2 \cdot \dfrac{9}{2} + y = 12,$ $9 + y = 12,$

$y = 3$

11. Solution:

$3y - \dfrac{1}{2}y = 25,$ $\dfrac{6y - y}{2} = 25$

$\dfrac{5y}{2} = 25$ $y = 10$

$x = 3y$

$x = 3 \cdot 10 = 30$

12. Solution:

$x = 6y$

$6y + 3y = 36,$ $9y = 36$

$y = 4$

$x = 6y,$ $x = 6 \cdot 4 = 24$

13. Solution:

$7k - 5 = -4k - 16$

$11k = -16 + 5,$ $11k = -11,$ $k = -1$

14. Solution:

$5x + 15 = 4x - 12$

$5x - 4x = -12 - 15,$ $x = -27$

15. Solution:

$3\ell + 2w = 48$

$w = 2\ell + 3$

$2\ell + 2(2\ell + 3) = 48$

$2\ell + 4\ell + 6 = 48$

$\ell = 7$

16. Solution:

$4C + 2A = 48$ $4C + 2(3C) = 48$

$C = \dfrac{1}{3}A,$ $A = 3C$ $4C + 6C = 48$

$10C = 48$

$C = \$4.8$

American Math Academy

1. Solution:

$(x - 7) \cdot (x + 7)$

2. Solution:

$$x^2 - 100$$

x   x   10   10

$(x - 10) \cdot (x + 10)$

3. Solution:

$$x^2 - 8x + 15$$

x   x   −3   −5

$(x - 3) \cdot (x - 5)$

4. Solution:

$$x^2 - x + 42$$

x   x   +6   −7

$(x - 7) \cdot (x + 6)$

5. Solution:

$$6x^2 - 21x - 12$$

6x   x   +3   −4

$(6x + 3) \cdot (x - 4)$

6. Solution:

$$x^2 - 2$$

x   x $\sqrt{2}$ $\sqrt{2}$

$(x - \sqrt{2}) \cdot (x + \sqrt{2})$

7. Solution:

$$\frac{x^2 \cdot x^2}{y^2 \cdot x^2} + \frac{y^2 \cdot y^2}{x^2 \cdot y^2} = \frac{x^4}{x^2 y^2} + \frac{y^4}{x^2 y^2} = \frac{x^4 + y^4}{x^2 y^2}$$

8. Solution:

$$x^2 + 2xy + y^2$$

x   x   y   y

$(x + y) \cdot (x + y)$

9. Solution:

$$x^2 - 6x - 16$$

x   x   +2 −8

$(x + 2) \cdot (x - 8)$

10. Solution:

$$x^2 - 9x + 8$$

x   x   −8 −1

$(x - 8) \cdot (x - 1)$

11. Solution:

$$7x^2 + 55x - 72$$

7x   x   −8   9

$(7x - 8) \cdot (x + 9)$

12. Solution:

$$2x^2 - 7x - 15$$

2x   x   +3 −5

$(2x + 3) \cdot (x - 5)$

13. Solution:

$$25x^2 + 5x - 2$$

5x 5x   +2 −1

$(5x - 1) \cdot (5x + 2)$

14. Solution:

Not factorable

15. Solution:

$(x + 3) \cdot (x + 1) = x^2 + x + 3x + 3$

$= x^2 + 4x + 3$

$k = 4$

American Math Academy

SOLVING QUADRATIC EQUATIONS BY FORMULA AND COMPLETE SQUARE EXERCISES SOLUTION

1. Solution:

$ax^2 + bx + c = 0$

$a = 1$

$b = 8$

$c = 12$

$x = \dfrac{-8 \mp \sqrt{64 - 4(1)(12)}}{2 \cdot 1} = \dfrac{-8 \mp \sqrt{64 - 48}}{2}$

$= \dfrac{-8 \mp \sqrt{16}}{2}$

$x = \dfrac{-8 \mp 4}{2}, x = -6, x = -2$

3. Solution:

$ax^2 + bx + c = 0$

$a = 4$

$b = 5$

$c = 1$

$x = \dfrac{-5 \mp \sqrt{25 - 4(4) \cdot (1)}}{2 \cdot 4} = \dfrac{-5 \mp \sqrt{25 - 16}}{8}$

$= \dfrac{-5 \mp \sqrt{9}}{8}$

$x = \dfrac{-5 \mp 3}{8}, x = -1, x = \dfrac{-1}{4}$

5. Solution:

$ax^2 + bx + c = 0$

$a = 1$

$b = -1$

$c = -1$

$x = \dfrac{1 \mp \sqrt{(-1)^2 - 4(1)(-1)}}{2 \cdot 1} = \dfrac{1 \mp \sqrt{1 + 4}}{2} = \dfrac{1 \mp \sqrt{5}}{2}$

2. Solution:

$ax^2 + bx + c = 0$

$a = 1, \qquad b = 2, \qquad c = 1$

$x = \dfrac{-2 \mp \sqrt{2^2 - 4(1)(1)}}{2} = \dfrac{-2 \mp \sqrt{4 - 4}}{2}$

$= \dfrac{-2}{2} = -1$

4. Solution:

$ax^2 + bx + c = 0$

$a = 1 \qquad b = -8 \qquad c = 12$

$x = \dfrac{8 \mp \sqrt{64 - 4(1)(12)}}{2 \cdot 4} = \dfrac{8 \mp \sqrt{64 - 48}}{2}$

$x = \dfrac{8 \mp \sqrt{16}}{2} = \dfrac{8 \mp 4}{2}, x = 6, x = 2$

6. Solution:

$ax^2 + bx + c = 0$

$a = 1 \qquad b = -4 \qquad c = 3$

$x = \dfrac{4 \mp \sqrt{(-4)^2 - 4(1)(3)}}{2 \cdot 1} = \dfrac{4 \mp \sqrt{16 - 12}}{2}$

$x = \dfrac{4 \mp \sqrt{4}}{2} = \dfrac{4 \mp 2}{2} = 2 \mp 1$

$x = 3 \text{ or } x = 1$

American Math Academy

7. Solution:

$ax^2 + bx + c = 0$

$a = 1 \qquad b = -6 \qquad c = 1$

$x = \dfrac{6 \mp \sqrt{(-6)^2 - 4(1)(1)}}{2 \cdot 1} = \dfrac{6 \mp \sqrt{36 - 4}}{2}$

$= \dfrac{6 \pm \sqrt{32}}{2}$

$x = \dfrac{6 \mp 4\sqrt{2}}{2} = 3, \mp 2\sqrt{2}$

8. Solution:

$ax^2 + bx + c = 0$

$a = 3, \qquad b = 4, \qquad c = -3$

$x = \dfrac{-4 \mp \sqrt{(-4)^2 - 4(3) \cdot (-3)}}{2 \cdot 3} = \dfrac{-4 \mp \sqrt{16 + 36}}{6}$

$x = \dfrac{-4 \mp \sqrt{52}}{6}$

$x = \dfrac{-4 \mp 2\sqrt{13}}{6}$

$x = \dfrac{-2 \mp 2\sqrt{13}}{3}$

9. Solution:

$x^2 + 6x - 7 = 0$

$(x + 3)^2 - 9 - 7 = 0$

$(x + 3)^2 - 16 = 0 \quad , \quad (x + 3)^2 = 16$

$x + 3 = \mp 4 \qquad\qquad x = -3 \mp 4$

$x = -7 \text{ or } x = 1$

10. Solution:

$(x + 3)^2 - 9 - 16 = 0$

$(x + 3)^2 = 25, \ x + 3 = \mp 5$

$x = 2 , \ x = -8$

12. Solution:

$(x - 3)^2 - 9 - 20 = 0$

$(x - 3)^2 = 29$

$(x - 3) = \mp \sqrt{29}$

$x = 3 \mp \sqrt{29}$

11. Solution:

$(x^2 + 12x) + 8 = (x + 6)^2 - 36 + 8 = 0$

$(x + 6)^2 = 28 \qquad x + 6 = \mp\sqrt{28}$

$x = -6 \mp 2\sqrt{7}$

13. Solution:

$(2x^2 + 8x) - 10 = 0,$ $2(x^2 + 4x) - 10 = 0$

$2(x + 2)^2 - 8 - 10 = 0$ $2(x + 2)^2 = 18$

$(x + 2)^2 = 9$

$x + 2 = \sqrt{9}$

$x + 2 = \mp 3$, $x + 2 = 3$ or $x + 2 = -3$

$x = 1$ or $x = -5$

14. Solution:

$3(x^2 + 2x) - 18 = 0,$ $3(x + 1)^2 - 3 - 18 = 0$

$3(x + 1)^2 = 21$ $(x + 1)^2 = 7$

$x + 1 = \mp \sqrt{7}$ $x = -1 \mp \sqrt{7}$

15. Solution:

$5(x^2 + 4x) - 40 = 0$

$5(x + 2)^2 - 20 - 40 = 0$

$5(x + 2)^2 = 60,$ $(x + 2)^2 = 12$

$x + 2 = \mp \sqrt{12}$

$x = -2 \mp 2\sqrt{3}$

16. Solution:

$(x^2 + 4x) + 3 = 0$

$(x + 2)^2 - 4 + 3 = 0,$ $(x + 2)^2 = 1$

$x + 2 = \mp 1$

$x = -2 \mp 1$

$x = -3, -1$

American Math Academy

1. Solution:

$= 5x^2 + 12x + 12x - 9 = 5x^2 + 24x - 9$

2. Solution:

$= -7x^2 + 10x + 8x - 11 = -7x^2 + 18x - 11$

3. Solution:

$= -x^2 + 12x + 8 + 5x - 8$

$= -x^2 + 17x$

4. Solution:

$= 6x^2 + x - 8x - 13 = 6x^2 - 7x - 13$

5. Solution:

$= 12x^2 - 12x - 12x - 4$

$= 12x^2 - 24x - 4$

6. Solution:

$= 2x^2 + 4x + 4 + 2x - 18$

$= 2x^2 + 9x - 14$

7. Solution:

$= x^2 + 8x + 6 - 2x + 18$

$= x^2 + 6x + 24$

8. Solution:

$= -x^2 - 10x + 5x + 13 = -x^2 - 5x + 13$

9. Solution:

$= x^2 - x + x^2 + 8x - 9$

$= 2x^2 + 7x - 9$

10. Solution:

$= 4x^2 - 3x - 8x^2 + 8x = -4x^2 + 5x$

11. Solution:

$= x^4 - 6x^3 - 12x^2 - 5x^2 + 30x + 60$

$= x^4 - 6x^3 - 17x^2 + 30x + 60$

12. Solution:

$= 7x^2 - x - x^2 + 14x - 4$

$= 6x^2 + 13x - 4$

13. Solution:

$= x^4 - 8x^3 + ax^2 + x^3 + 8x^2 - 9x$

$= -x^4 - 7x^3 + 17x^2 - 9x$

14. Solution:

$= 4x^2 - 3x - 8x^2 - 8x$

$-4x^2 - 11x$

15. Solution:

$= -5x^2 - 18x + 9 - x^2 + 8x + 4$

$= -6x^2 - 10x + 13$

16. Solution:

$= 4x^2 - 3x + 8x$

$= 4x^2 + 5x$

MULTIPLYING AND DIVIDING POLYNOMIALS EXERCISES SOLUTIONS

1. Solution:

$= (a^{12}b^4) \cdot (a^1b^5)$

$= a^{13}b^9$

2. Solution:

$= a^3 \cdot a^8 \cdot b^7 \cdot b^{10}$

$= a^{11} \, b^{17}$

3. Solution:

$= a^{12} \cdot (a^6 \, b^3)$

$= a^{18} \, b^3$

4. Solution:

$= 6a^3 \cdot a^4 \cdot b^8$

$= 6a^7 \, b^8$

5. Solution:

$= (a^{12} \, b^8)(a^1 \, b^{15}) = a^{13} \, b^{23}$

6. Solution:

$= a^{13} \cdot a^8 \cdot b^{17} \, b^{10} = a^{21} \, b^{27}$

7. Solution:

$= 2a^6 - a^5 - 6a^5 + 3a^4$

$= 2a^6 - 7a^5 + 3a^4$

8. Solution:

$= a^2 + a + 2 + a^2 + 5a$

$= 2a^2 + 6a + 2$

9. Solution:

$= 34 \, a^{10}$

10. Solution:

$= 3a2(a3) - 3a^2(7) + 3a^2(7)$

$= 3a^5 - 3a^3 + 21a^2$

11. Solution:

$= \dfrac{\cancel{5}(3a-5)}{\cancel{5}} = 3a - 5$

12. Solution:

$= \dfrac{\cancel{6}(a^3+2a^2-10)}{\cancel{6}} = a^3 + 2a^2 - 10$

13. Solution:

$= \dfrac{\cancel{a}(a^3+5a^2+9)}{\cancel{a}}$

$= a^3 + 5a^2 + 9$

14. Solution:

$= 3(2x^2 + x - 8x - 4)$

$= 3(2x^2 - 7x - 4) = 6x^2 - 21x - 12$

15. Solution:

$= \dfrac{(a^2-b^2)\cdot(a^2+ab)}{(a^2+ab)\cdot(ab+a)} = \dfrac{(a-b)\cancel{(a+b)}}{\cancel{a(a+b)}} \cdot \dfrac{\cancel{a}(a+b)}{a(b+1)}$

$= \dfrac{(a-b)\cdot(a+b)}{a(b+1)} = \dfrac{a^2-b^2}{ab+a}$

16. Solution:

$= \dfrac{(a-b)\cancel{(a+b)}}{a\cancel{(a+b)}} \, \dfrac{\cancel{a}(b+a)}{\cancel{a}(a+b)} = \dfrac{a-b}{a}$

American Math Academy

1. Solution:

$5(2x - 3) = 4x$

$10x - 15 = 4x,$ $6x = 15$

$x = \dfrac{15}{6} = \dfrac{5}{2}$

2. Solution:

$5(x - 3) = 2(x - 5)$

$5x - 15 = 2x - 10$

$3x = 5$

$x = \dfrac{5}{3}$

3. Solution:

$\dfrac{3x - 3 + 4x}{\overset{12}{\underset{6}{}}} = \dfrac{1}{2}$

$3x - 3 + 4x = 6$

$7x = 9$

$x = \dfrac{9}{7}$

4. Solution:

$\dfrac{6x + 15 + x}{\overset{18}{\underset{2}{}}} = \dfrac{1}{\overset{9}{\underset{1}{}}}$

$\dfrac{7x + 15}{2} = 1$

$7x + 15 = 2$

$7x = -13$

$x = \dfrac{-13}{7}$

5. Solution:

$\dfrac{2x + x}{8} = 7$

$\dfrac{3x}{8} = 7$

$3x = 56$

$x = \dfrac{56}{3}$

6. Solution:

$\dfrac{4x - x}{24} = 48$

$\dfrac{3x}{\underset{8}{24}} = 48$

$x = 8 \cdot 48$

$x = 384$

7. Solution:

$\dfrac{2x - 2 + x}{8} = \dfrac{7x}{2}$

$\dfrac{3x - 2}{\underset{4}{8}} = \dfrac{7x}{\underset{1}{2}}$

$3x - 2 = 28x$

$-2 = 25x$

$\dfrac{-2}{25} = x$

8. Solution:

$\dfrac{7x - 7 + x + 1}{14} = 5$

$\dfrac{8x - 6}{14} = 5$

$8x - 6 = 70$

$8x = 76$

$x = \dfrac{19}{2}$

9. Solution:

$$\frac{x}{4} - \frac{x}{2} \cdot \frac{2}{2} = 5$$

$$\frac{x}{4} - \frac{2x}{4} = 5$$

$$\frac{-x}{4} = 5$$

$$x = -20$$

11. Solution:

$$2x - 24 = 4x - 36$$

$$36 - 24 = 2x$$

$$12 = 2x$$

$$6 = x$$

13. Solution:

$$= \frac{1}{x} \cdot \frac{x-1}{3} = \frac{1}{4}$$

$$4x - 4 = 3x$$

$$x = 4$$

15. Solution:

$$\frac{x-2}{3} \cdot \frac{4}{x-5} = \frac{1}{12}$$

$$\frac{4x-8}{3x-15} = \frac{1}{12}$$

$$48x - 96 = 3x - 15$$

$$45x = 81$$

$$x = \frac{81}{45} = \frac{9}{5}$$

17. Solution:

$$\frac{(x+6)(x-6)}{x-6} = 12$$

$$x + 6 = 12$$

$$x = 6$$

10. Solution:

$$7 = \frac{x}{3} \cdot \frac{2}{2} - \frac{x}{6}$$

$$7 = \frac{2x}{6} - \frac{x}{6}$$

$$7 = \frac{x}{6}$$

$$42 = x$$

12. Solution:

$$= \frac{2x+5}{36} = \frac{1}{6}$$

$$(2x + 5) \cdot \overset{1}{6} = \overset{6}{36}$$

$$2x = 1$$

$$x = \frac{1}{2}$$

14. Solution:

$$\frac{12+x}{12x} = \frac{1}{36}, \qquad \frac{12+x}{x} = \frac{1}{3}$$

$$3(12 + x) = x \qquad 36 + 3x = x$$

$$36 = -2x \qquad -18 = x$$

16. Solution:

$$\frac{(x-2)(x+2)}{(x-2)} = 6$$

$$x + 2 = 6 \qquad x = 4$$

18. Solution:

$$\frac{2+1}{2x} = \frac{1}{36}$$

$$\frac{3}{2x} = \frac{1}{36}$$

$$x = 3 \cdot 18$$

$$x = 54$$

1. Solution:

$$= \frac{x-2}{(x+2)(x-2)} + \frac{x+2}{(x-2)(x+2)}$$

$$= \frac{x-2+x+2}{(x-2)(x+2)} = \frac{2x}{(x-2)(x+2)}$$

2. Solution:

$$= \frac{x+4-1(-x-4)}{(-x-4)(x+4)} = \frac{x+4+x+4}{(-x-4)(x+4)}$$

$$= \frac{2x+8}{(-x-4)(x+4)} = \frac{2(x+4)}{(x+4)(-x-4)} = \frac{2}{-x-4}$$

3. Solution:

$$= \frac{x(x+1)}{(x-1)(x+1)} + \frac{2x(x-1)}{(x+1)(x-1)}$$

$$= \frac{x^2+x+2x^2-2x}{(x-1)(x+1)} = \frac{3x^2-x}{(x-1)(x+1)}$$

4. Solution:

$$\frac{1(x+4)-3(1)}{3(x+4)} = \frac{x+4-3}{3(x+4)}$$

$$= \frac{x+1}{3(x+4)}$$

5. Solution:

$$= \frac{2x(x+3)}{(x-1)(x+3)} - \frac{5(x-1)}{(x+3)(x-1)}$$

$$= \frac{2x^2+6x-5x+5}{(x+3)(x-1)} = \frac{2x^2+x+5}{(x+3)(x-1)}$$

6. Solution:

$$= \frac{x-1+3x}{3x(x-1)} = \frac{4x-1}{3x(x-1)}$$

7. Solution:

$$\frac{2x^2}{12x^6} = \frac{1}{6}x^{2-6} = \frac{1}{6}x^{-4} = \frac{1}{6x^4}$$

8. Solution:

$$\frac{3x^{-4}}{9x^8} = \frac{1}{3}x^{-4-8} = \frac{1}{3}x^{-12} = \frac{1}{3x^{12}}$$

9. Solution:

$$\frac{\cancel{(x-6)}(x+6)}{\cancel{(x-6)}} = x+6$$

10. Solution:

$$\frac{1}{3}x^{-4-12}y^{-7+5}$$

$$= \frac{1}{3}x^{-16} \cdot y^{-2} = \frac{1}{3x^{16}y^2}$$

11. Solution:

$$\frac{x(x+3)}{3x} = \frac{x+3}{3} = \frac{x}{3}+1$$

12. Solution:

$$= x^{-1-10} \cdot y^{-3-4} \cdot z^{4-(-2)}$$

$$= x^{-11} y^{-7} \cdot z^6 = \frac{z^6}{x^{11}y^7}$$

13. Solution:

$$= \frac{\cancel{(x-y)}(x+y)}{\underset{1}{\cancel{3x}}} \cdot \frac{9x}{\cancel{x-y}} = \frac{(x+y) \cdot 3}{1} = 3x+3y$$

14. Solution:

$$\frac{\overset{x}{\cancel{x^2}}}{\underset{1}{\cancel{3x}}} \cdot \frac{\overset{1}{\cancel{9x}}}{\underset{4}{\cancel{36}}} = \frac{x}{3} \cdot \frac{x}{4} = \frac{x^2}{12}$$

15. Solution:

$$\frac{(x-y)\cancel{(x+y)}}{x} \cdot \frac{x}{\cancel{(x+y)}}$$

$$= x-y$$

16. Solution:

$$\frac{x^2}{\underset{1}{\cancel{3}}} \cdot \frac{\overset{6}{\cancel{18}}}{x^8} = 6x^{2-8} = 6x^{-6} = \frac{6}{x^6}$$

17. Solution:

$$\frac{5x\cancel{(x-1)}}{3x} \cdot \frac{1}{\cancel{x-1}} = \frac{5}{3}$$

18. Solution:

$$\frac{(x-y)\cancel{(x^2+xy+y^2)}}{\cancel{(x^2+xy+y^2)}} = x-y$$

1. Solution:

Domain: 2, 5, 6

Yes, it is a function because domain elements do not repeat.

2. Solution:

Domain: (1, 2, 3, 1)

No, it is a not function because domain elements are repeated.

3. Solution:

Domain: e, c, m

Yes it is a function because domain elements do not repeat.

4. Solution:

Domain: a, b, c

Yes, it is a function because domain elements do not repeat.

5. Solution:

Domain: m, n, k

Yes, it is a function because domain elements do not repeat.

6. Solution:

No, it is a not function because domain elements are repeated.

7. Solution:

$f(x) = y$

$y = x + 4$, $y - 4 = x$ (change x to y)

$f^{-1}(x) = x - 4$

8. Solution:

$f(x) = y$

$y = 2x - 54$, $y + 54 = 2x$, $\dfrac{y + 54}{2} = x$

change (x to y), $f^{-1}(x) = \dfrac{x + 54}{2}$

9. Solution:

$f(x) = y$

$y = x^2 - 3$, $y + 3 = x^2$

$\sqrt{y + 3} = x$ (change x to y)

$f^{-1}(x) = \sqrt{x + 3}$

10. Solution:

$f(x) = y$

$(y)^2 = \left(\sqrt{x - 6}\right)^2$

$y^2 = x - 6$

$y^2 = x - 6$

$y^2 + 6 = x$ (change x to y)

$f^{-1}(x) = x^2 + 6$

American Math Academy

FUNCTION OPERATIONS
EXERCISES SOLUTIONS

1. Solution:

$= -3x + 9 + 2x + 5$

$= -x + 14$

2. Solution:

$= -3x + 9 - (2x + 5)$

$= -3x + 9 - 2x - 5 = -5x + 4$

3. Solution:

$= x^2 + 3x + 9 + 2(x + 12)$

$= x^2 + 3x + 9 + 2x + 24$

$= x^2 + 5x + 33$

4. Solution:

$= 3(x^2 + 3x + 9) - (x + 12)$

$= 3x^2 + 9x + 27 - x - 12$

$= 3x^2 + 8x + 15$

5. Solution:

$= (3x + 10) + 4$

$= 3x + 14$

6. Solution:

$= 3(x + 4) + 10$

$= 3x + 12 + 10 = 3x + 22$

7. Solution:

$= 5(6x - 15)$

$= 30x - 75$

8. Solution:

$= 6(5x) - 15$

$= 30x - 15$

9. Solution:

$= g(3) = 3 - 3 = 0$

$= f(0) = 2 \cdot 0^2 + 1 = 1$

10. Solution:

$= f(5) = 2 \cdot 5^2 + 1 = 50 + 1 = 51$

$= g(51) = 51 - 3 = 48$

11. Solution:

$= g(-1) = 4(-1)^2 + 5(-1) - 12 = 4 - 5 - 12 = -13$

$= f(-13) = 2(-13)^2 + (-13) + 3 = 2 \cdot 169 - 13 + 3 = 328$

12. Solution:

$f(-3) = 2(-3)^2 + (-3) + 3$

$= 2 \cdot 9 + 0$

$= 18$

$g(18) = 4(18)^2 + 5 \cdot (18) - 12$

$= 1296 + 90 - 12 = 1374$

13. Solution:

$= (x^2 - x)(x^2 + x - 2)$

$= (x^4 + x^3 - 2x^2 - x^3 - x^2 + 2x)$

$= x^4 - 3x^2 + 2x$

14. Solution:

$= \dfrac{x^2 - x}{x^2 + x - 2} = \dfrac{x(x-1)}{(x+2)(x-1)}$

$= \dfrac{x}{x+2}$

MEAN, MEDIAN AND MODE
EXERCISE SOLUTIONS

1.

Mean: $\dfrac{3+4+5+6+6+7+11}{7} = 6$

Median: 3, 4, 5, ⑥, 6, 7, 11 = 6

Mode: Most repeated number: 6

Range: 11 − 3 = 8

4. $\dfrac{y+6}{2} = 10, y+6 = 20$

$\qquad\qquad\qquad y = 14$

2.

Mean: $\dfrac{1+3+5+7+7+1}{6} = 5$

Median: 1, 3, 5, 7, 7, 7 $= \dfrac{5+1}{2} = ⑥$

Mode: Most repeated number: 7

Range: 7 − 1 = 6

5. $\dfrac{z+12}{2} = 20, z+12 = 40$

$\qquad\qquad\qquad z = 28$

3.

Mean: $\dfrac{\frac{1}{2}+\frac{1}{3}+\frac{1}{4}+\frac{1}{5}}{4}$

$= \dfrac{\frac{30+20+15+12}{60}}{4}$

$= \dfrac{77}{240} = 0 \cdot 32$

Median: $\dfrac{\frac{1}{3}+\frac{1}{4}}{2} = \dfrac{4+3}{24} = \dfrac{7}{24}$

Mode: Most repeated number: No mode

Range: $\dfrac{1}{2} - \dfrac{1}{5} = \dfrac{5-2}{10} = \dfrac{3}{10}$

6. $\dfrac{12 \cdot 5 + k}{2} = 30 \cdot 5$

$12 \cdot 5 + k = 61$

$\qquad\qquad k = 48 \cdot 5$

7. $\dfrac{x+x+1+x+2+x+3+x+4}{5} = 30$

$\dfrac{5x+10}{5} = 30$

$x + 2 = 30,$

$x = 28,$

Greatest possible value = x + 4 = 28 + 4

$= 32$

8. Range: Difference befween the lowest and highest values

136

AMERICAN MATH
—ACADEMY—

SLOPE AND SLOPE INTERCEPT
FORM EXERCISE SOLUTIONS

1. Solution:

$m = \dfrac{9-4}{7-3} = \dfrac{5}{4}$

2. Solution:

$m = \dfrac{8-1}{6-2} = \dfrac{7}{4}$

3. Solution:

$m = \dfrac{7-9}{3-4} = \dfrac{-2}{-1} = 2$

4. Solution:

$\dfrac{-7-6}{-1-(-3)} = \dfrac{-13}{2}$

5. Solution:

$m = \dfrac{8-2}{1-3} = \dfrac{6}{-2} = -3$

6. Solution:

$m = \dfrac{2-3}{3-0} = \dfrac{-1}{3}$

7. Solution:

$m = \dfrac{\frac{-1}{3} - \frac{1}{3}}{1-(-2)} = \dfrac{\frac{-2}{3}}{3} = \dfrac{-2}{9}$

8. Solution:

$m = \dfrac{\frac{1}{4} - \frac{1}{2}}{6-2} = \dfrac{-\frac{1}{4}}{4}$

$m = -\dfrac{1}{16}$

9. Solution:

$y = 3x + b$
$y = 3x - 19$

10. Solution:

$y = 6x + b,$ $3 = 6(-1) + b$
                                    $9 = b$
                           $y = 6x + 9$

11. Solution:

$y = 5x + b$
$y = 5x + 6$

12. Solution:

$y = -2x + b,$ $-6 = -2(-4) + b$
$b = -14$
$y = -2x - 14$

13. Solution:

$y = \dfrac{-1}{3}x + b,$ $y = -\dfrac{1}{3}x + 6$

14. Solution:

$y = \dfrac{-2}{5}x + b,$ $y = \dfrac{-2}{5}x + \dfrac{1}{5}$

15. Solution:

$y = mx + b$
$m = \dfrac{1}{4}$

16. Solution:

$y = mx + b$
$m = \dfrac{-1}{5}$

17. Solution:

$y = mx + b$
$m = -7$

18. Solution:

$y = mx + b$
$m = 8$

DISTANCE AND MIDPOINT EXERCISE SOLUTIONS

1. Solution:

$d = \sqrt{(8-5)^2 + (7-3)^2}$

$d = \sqrt{9+16} = \sqrt{25}$

$d = 5$

2. Solution:

$d = \sqrt{(-8-3)^2 + (-5-(-1))^2} = \sqrt{(-11)^2 + (-4)^2}$

$d = \sqrt{121+16} = \sqrt{137}$

3. Solution:

$d = \sqrt{(12-4)^2 + (-3-6)^2} = \sqrt{64+81} = \sqrt{145}$

4. Solution:

$d = \sqrt{(6-6)^2 + \left(\dfrac{1}{2}-\dfrac{1}{3}\right)^2} = \sqrt{0 + \left(\dfrac{1}{6}\right)^2}$

$d = \sqrt{\dfrac{1}{36}} = \dfrac{1}{6}$

5. Solution:

$d = \sqrt{(8-2)^2 + \left(\dfrac{1}{2}-\dfrac{1}{2}\right)^2} = \sqrt{36+0}$

$d = 6$

6. Solution:

$d = \sqrt{(0-0)^2 + \left(\dfrac{-1}{4}-\dfrac{1}{2}\right)^2} = \sqrt{\left(\dfrac{-1-2}{4}\right)^2} = \sqrt{\left(\dfrac{-3}{4}\right)^2}$

$d = \sqrt{\dfrac{9}{16}} = \dfrac{3}{4}$

7. Solution:

$d = \sqrt{(3-7)^2 + (7-6)^2} = \sqrt{16+1} = \sqrt{17}$

8. Solution:

$d = \sqrt{(8-14)^2 + (2-10)^2} = \sqrt{36+64}$

$d = \sqrt{100} = 10$

9. Solution:

$d = \sqrt{(-3-8)^2 + (-1-6)^2} = \sqrt{(-11)^2 + (-7)^2}$

$d = \sqrt{121+49} = \sqrt{170}$

10. Solution:

$d = \sqrt{(-1+3)^2 + (-4-9)^2} = \sqrt{4+169}$

$d = \sqrt{173}$

Academy
American Math Academy

138 AMERICAN MATH ACADEMY

DISTANCE AND MIDPOINT EXERCISE SOLUTIONS

11. Solution:

$$M_x = \frac{0+2}{2} = 1$$
$$M_y = \frac{9+5}{2} = 7 \left.\right\} (1,7)$$

12. Solution:

$$M_x = \frac{-2-8}{2} = -5$$
$$M_y = \frac{-8+4}{2} = -2 \left.\right\} (-5,-2)$$

13. Solution:

$$M_x = \frac{7-1}{2} = 3$$
$$M_y = \frac{6+0}{2} = 3 \left.\right\} (3,3)$$

14. Solution:

$$M_x = \frac{\frac{1}{3}+\frac{1}{3}}{2} = \frac{1}{3}$$
$$M_y = \frac{9+9}{2} = 9 \left.\right\} \left(\frac{1}{3},9\right)$$

15. Solution:

$$M_x = \frac{\frac{1}{6}-\frac{1}{6}}{2} = 0$$
$$M_y = \frac{7+1}{2} = 4 \left.\right\} (0,4)$$

16. Solution:

$$M_x = \frac{\frac{1}{8}+\frac{1}{8}}{2} = \frac{1}{8}$$
$$M_y = 0 \left.\right\} \left(\frac{1}{8},0\right)$$

17. Solution:

$$M_x = \frac{9+7}{2} = 8$$
$$M_y = \frac{3+7}{2} = 5 \left.\right\} (8,5)$$

18. Solution:

$$M_x = \frac{-1+9}{2} = 4$$
$$M_y = \frac{23-13}{2} = 5 \left.\right\} (4,5)$$

American Math Academy

1. **Solution:**

   $a^2 + b^2 = c^2$

   $6^2 + 8^2 = x^2$

   $36 + 64 = x^2$

   $\sqrt{100} = \sqrt{x^2}$

   $10 = x$

2. **Solution:**

   $a^2 + b^2 = c^2$

   $9^2 + 12^2 = x^2$

   $81 + 144 = x^2$

   $\sqrt{225} = \sqrt{x^2}$

   $15 = x$

3. **Solution:**

   $a^2 + b^2 = c^2$

   $(2\sqrt{2})^2 + (4\sqrt{2})^2 = x^2$

   $8 + 32 = x^2$

   $\sqrt{40} = \sqrt{x^2}$

   $x = 2\sqrt{10}$

4. **Solution:**

   $a^2 + b^2 = c^2$

   $x^2 + 12^2 = 15^2$

   $x^2 + 144 = 225$

   $\sqrt{x^2} = \sqrt{81}$

   $x = 9$

5. **Solution:**

   $a^2 + b^2 = c^2$

   $x^2 + 20^2 = 25^2$

   $x^2 + 400 = 625$

   $\sqrt{x^2} = \sqrt{225}$

   $x = 15$

6. **Solution:**

   $a^2 + b^2 = c^2$

   $x^2 = 6^2 + 8^2$

   $x^2 = 100$

   $x = 10$

   $3^2 + 4^2 = y^2$

   $\sqrt{25} = \sqrt{y^2}$

   $y = 5$

7. **Solution:**

   $a^2 + b^2 = c^2$

   $x^2 + 20^2 = 25^2$

   $x^2 + 400 = 625$

   $\sqrt{x^2} = \sqrt{225}$

   $x = 15$

8. **Solution:**

   $a^2 + b^2 = c^2$

   $3^2 + (BD)^2 = 25$

   $(BD)^2 = 16,$　　　　$BD = 4$

   $3^2 + 8^2 = x^2$

   $9 + 64 = x^2$

   $\sqrt{73} = \sqrt{x^2},$　　　$x = \sqrt{73}$

American Math Academy

1. $27^x = 81^y$

$3^{3x} = 3^{4y}$

$3x = 4y.$

Ratio of y to x = $\dfrac{3}{4}$.

Correct Answer : B

2. Three consecutive integers = x, x + 1, x + 2.

$x + x + 1 + x + 2 = 96$

$3x + 3 = 96$

$3x = 93$

$x = 31$

Highest one = x + 2 = 31 + 2 = 33.

Correct Answer : C

3. $= 2(y - 6) = 3(2y - 12)$

$2y - 12 = 6y - 36$

$-12 + 36 = 6y - 2y$

$24 = 4y$

$6 = y$

Correct Answer : B

4. $\dfrac{x+4}{3} - \dfrac{x}{5} = \dfrac{14}{15}$

$= \dfrac{5(x+4)}{3.5} - \dfrac{3(x)}{3.5} = \dfrac{14}{15}$

$5x + 20 - 3x = 14$

$2x = 14 - 20$

$2x = -6$

$x = -3$

Correct Answer : C

5. $x = \dfrac{5^3}{\sqrt{125}}$

$x = \dfrac{125}{5\sqrt{5}}$

$x = \dfrac{25}{\sqrt{5}}\left(\dfrac{\sqrt{5}}{\sqrt{5}}\right)$

$x = \dfrac{25\sqrt{5}}{5} = 5\sqrt{5}$

Correct Answer : A

6. $\dfrac{x^2 - 25}{x+5} = \dfrac{(x-5)(x+5)}{x+5} = x - 5.$

Correct Answer : D

American Math Academy

—ACADEMY—

141

7. $=\sqrt{32} - \sqrt{8} + \sqrt{128}$

$= 4\sqrt{2} - 2\sqrt{2} + 8\sqrt{2}$

$= 10\sqrt{2}$

Correct Answer : D

10. $= \sqrt{-8} \cdot \sqrt{32}$

$= \sqrt{8i^2} \cdot \sqrt{32}$

$= 2i\sqrt{2} \cdot 4\sqrt{2}$

$= 8i \cdot 2$

$= 16i$

Correct Answer : C

8. $\dfrac{x^2 - y^2}{x^2 - xy} \cdot \dfrac{xy - x}{x^2 + xy}$

$= \dfrac{\cancel{(x-y)}(x+y)}{x\cancel{(x-y)}} \cdot \dfrac{x(y-1)}{x(x+y)}$

$= \dfrac{y-1}{x}$

Correct Answer : C

11. $\dfrac{3+i}{2-i} = \dfrac{(3+i)(2+i)}{(2-i)(2+i)}$

$= \dfrac{6 + 3i + 2i + i^2}{4 + 2i - 2i - i^2}$

$= \dfrac{5 + 5i}{5}$

$= 1 + i \qquad\qquad i^2 = -1$

Correct Answer : A

9. $y = 3x^2 + 6x - 9$

$y = 3(x^2 + 2x) + 9$

$y = 3(x + 1)^2 - 3 - 9$

$y = 3(x + 1)^2 - 12$

Correct Answer : A

12. $2x - 18 = 5x - 14$

$-18 + 14 = 5x - 2x$

$-4 = 3x$

$-\dfrac{4}{3} = x$

Correct Answer : D

AMERICAN MATH
ACADEMY

American Math Academy

13. $\dfrac{a^{2m}}{a^{10}} = a^6$ and $a^{3n} = a^{30}$

$a^{2m} = a^{16}$, then m = 8

$a^{3n} = a^{30}$, then n = 10

m . n = 8 . 10 = 80

Correct Answer : C

14. $2^a \cdot 2^a \cdot 2^a \cdot 2^a = 8^{4b}$

$2^{a+a+a+a} = 2^{12b}$

4a = 12b

a = 3b

Correct Answer : A

15. $= (3 + 4i) \cdot (3 - 4i)$

$= 9 - 12i + 12i - 16(i)^2$

$= 9 + 16$

$= 25$

Correct Answer : D

16. $x^2 - 4x + y = 14$

$5x - y = 6$

$+$ _____

$x^2 + x = 20$

$x^2 + x - 20 = 0$

(x + 5)(x − 4) = 0, then x = −5 or x = 4

Correct Answer : D

17. Since the function is a zero function, $x^2 - 169 = 0$

$x^2 - 169 = 0$

$x^2 = 169, \quad \sqrt{x^2} = \sqrt{169}$

$x = \pm 13$

Correct Answer : B

American Math Academy

1.

$$\frac{1}{2^{3x}} = \frac{1}{32^2}$$

$$\frac{1}{2^{3x}} = \frac{1}{2^{10}}$$

$$2^{3x} = 2^{10}$$

$$x = \frac{10}{3}$$

Correct Answer : D

2. Direct variation $y = kx$.

$18 = k \cdot 3$

$6 = k$

$y = kx$

$y = 6 \cdot 4$

$y = 24$

Correct Answer : B

3.

$$\frac{7}{a} = \frac{4}{b}$$

$7b = 4a$

$$\frac{a}{b} = \frac{7}{4}$$

Correct Answer : A

4. $\frac{1}{3}(3x - 9) + (x - 12) = ax + x + b$

$x - 3 + x - 12 = ax + x + b$

$2x - 15 = x(a + 1) + b$

$2 = a + 1$

$1 = a$

$-15 = b$

$a - b = 1 - (-15)$

$\qquad = 16$

Correct Answer : B

5. $\frac{3}{5}x + ny = 25$, $\text{Slope} = \frac{3}{5n} = \frac{1}{25}$.

$5n = 75$

$n = 15$

Correct Answer : D

6. $\text{Slope} = \frac{y_2 - y_1}{x_2 - x_1}$

$m = \frac{14 - 8}{10 - 4} = 1$

Correct Answer : A

American Math Academy

7. $3x + 7 = -2x - 8$

$3x + 2x = -7 - 8$

$5x = -15$

$x = -3$

Correct Answer : C

8. Original price = 100x.

20% discount from original price

$= \dfrac{20 \cdot 100x}{100} = 20x.$

Sales price = 100x − 20x = \$60

80x = \$60

$x = \dfrac{60}{80} = \dfrac{6}{8} = \dfrac{3}{4}.$

Original price $= 100x = 100 \cdot \dfrac{3}{4} = \dfrac{300}{4} = 75.$

Correct Answer : D

9. $x = \dfrac{2^3}{\sqrt{12}}$

$x = \dfrac{8}{2\sqrt{3}} = \dfrac{4}{\sqrt{3}} \cdot \dfrac{\sqrt{3}}{\sqrt{3}} = \dfrac{4\sqrt{3}}{3}.$

Correct Answer : C

10. $9^{2x-2} = 81^{2x-3}$

$3^{2(2x-2)} = 3^{4(2x-3)}$

$2(2x - 2) = 4(2x - 3)$

$4x - 4 = 8x - 12$

$-4 + 12 = 8x - 4x$

$8 = 4x$

$2 = x$

Correct Answer : D

11. $x^2 - 8x - 12 = 0$

$x^2 - 8x = 12$

$(x - 4)^2 - 16 = 12$

$(x - 4)^2 = 16 + 12$

$(x - 4)^2 = 28$

$\sqrt{(x-4)^2} = \sqrt{28}$

$x - 4 = \pm 2\sqrt{7}$

$x = 4 \pm 2\sqrt{7}$

Correct Answer : A

12. $i^{2020} + i^{2021} + i^{2022}$

$= (i^2)^{2010} + (i^2)^{2010} \cdot i + (i^2)^{2011}$

$= (-1)^{2010} + (-1)^{2010} \cdot i + (-1)^{2011}$

$= 1 + i - 1$

$= i$

Correct Answer : C

American Math Academy

13. $\frac{3}{4}x - \frac{1}{8}x = \frac{1}{12} + \frac{2}{3}$

$\frac{3.2}{4.2}x - \frac{1}{8}x = \frac{1}{12} + \frac{2.4}{3.4}$

$\frac{6x - x}{8} = \frac{1 + 8}{12}$

$\frac{5x}{8} = \frac{9}{12}$

$60x = 72$

$x = \frac{72}{60} = \frac{6}{5}$

Correct Answer : D

14. $\frac{x^4 - 16}{x^2 - 4} = \frac{(x^2 - 4)(x^2 + 4)}{x^2 - 4} = x^2 + 4$

Correct Answer : D

15. $a = 2\sqrt{3}$ then , $a^2 = 12$.

$3a = \sqrt{3x}$

$9a^2 = 3x$

$9.12 = 3x$

$108 = 3x$

$36 = x$

Correct Answer : C

16. Area of circle = πr^2

$\pi r^2 = 16\pi$

$r^2 = 16$

$r = 4$

Circumference = $2\pi r = 2\pi . 4 = 8\pi$

Correct Answer : B

17. $x^3 + 6x^2 + 9x = 0 \rightarrow$ since this is a zero functions,

$x(x^2 + 6x + 9) = 0$

$x(x + 3)^2 = 0$

$x = 0$ or $x = -3$

Correct Answer : D

American Math Academy

FINAL TEST SOLUTIONS

1. Solution:

$$\frac{3x+15-8}{7}=\frac{17-6+x}{5}$$

$$\frac{3x+7}{7}=\frac{11+x}{5} \text{ (Cross multiply)}$$

$5(3x + 7) = 7(11 + x)$

$15x + 35 = 77 + 7x$

$15x - 7x = 77 - 35$

$8x = 42$

$$x=\frac{42}{8}=\frac{21}{4}$$

Correct Answer : A

2. Solution:

Since x is a positive number and $x^2 - 6x + 11 = 2$.

$x^2 - 6x + 9 = 0$

$(x - 3)(x - 3) = 0$

$x - 3 = 0$, then $x = 3$

Correct Answer : A

3. Solution:

$F = \frac{9}{5} C + 32°$ since C = 15°

$F = \frac{9}{5}(15°) + 32°$

$F = 27° + 32°$

$F = 59°$

Correct Answer : D

4. Solution:

$$\frac{x^2-16}{x-4}=\frac{(x-4)(x+4)}{x-4}=x+4$$

Correct Answer : D

5. Solution:

$\frac{a^{2m}}{a^{10}}=a^{12}$ and $a^{3n} = a^{21}$.

$a^{2m} = a^{22}$, then m = 11.

$a^{3n} = a^{21}$, then n = 7

$m \cdot n = 11 \cdot 7 = 77$

Correct Answer : C

6. Solution:

$f(x) = x^3 + 8x^2 + 16x = 0$

$f(x) = x(x^2 + 8x + 16) = 0$

$f(x) = x(x + 4)(x + 4) = 0$

$x = 0$ or $x + 4 = 0$, $x = -4$

Correct Answer : C

7. Solution:

$$\frac{2x+4}{6}-\frac{x}{4}=\frac{3}{2}$$

$$\frac{4x+8}{12}-\frac{3x}{12}=\frac{18}{12} \text{ (cancel denominator)}$$

$4x + 8 - 3x = 18$

$x + 8 = 18$

$x = 10$

Correct Answer : A

American Math Academy

FINAL TEST SOLUTIONS

8. Solution:

$= 4 + (2+3) \cdot 4 - 8 + (5+3)^0$

$= 4 + (5) \cdot 4 - 8 + 1$

$= 4 + 20 - 8 + 1$

$= 17$

Correct Answer : C

9. Solution:

$\dfrac{3}{x+1} = \dfrac{1}{x-1}$

$3(x-1) = x+1$

$3x - 3 = x + 1$

$2x = 4$

$x = 2$

Correct Answer : A

10. Solution:

Factors of 48: 1, 2, 3, 4, 6, 8, 12, 16, 24, 48

Correct Answer : C

11. Solution:

Use prime factorization to find the greatest prime factor of 14 and 66.

If x is the greatest prime factor of 14, then 14 = 2 · 7 and the greatest prime factor of 14 is 7. (x = 7)

If y is the greatest prime factor of 66, then 66 = 6 . 11 and the greatest prime factor of 66 is 11. (y = 11)

x + y = 7 + 11 = 18

Correct Answer : C

12. Solution:

Inverse variation y · x = k. Only Choice B can be inverse variation

Correct Answer : B

13. Solution:

% decreased from original price =

$\dfrac{\$400.25}{100} = \$100$

Television sale price = $400 – $100 = $300

Correct Answer : D

14. Solution:

Range means the difference between the lowest and highest number values.

Correct Answer : D

15. Solution:

Mean $= \dfrac{64 + 72 + 54 + 34 + x}{5}$

$48 = \dfrac{x + 224}{5}$

$5 \cdot 48 = x + 224$

$240 - 224 = x$

$16 = x$

Correct Answer : D

American Math Academy

16. Solution:

$$\sqrt{(16)^2} + \sqrt{2^2} - (-2)^3 = 16 + 2 + 8 = 26$$

Correct Answer : D

17. Solution:

$$4^{x-4} = 8^{x-6}$$

$$2^{2x-8} = 2^{3x-18}$$

$$2x - 8 = 3x - 18$$

$$x = 18 - 8$$

$$x = 10$$

Correct Answer : B

18. $2k - 5 + 3k = -4(k + 4)$

$$5k - 5 = -4k - 16$$

$$5k + 4k = -16 + 5$$

$$9k = -11$$

$$k = -\frac{11}{9}$$

Correct Answer : A

19. Solution:

$$\frac{1}{2}(4x - 6) + 5 = 3x - 7$$

$$2x - 3 + 5 = 3x - 7$$

$$2x + 2 = 3x - 7$$

$$2 + 7 = x$$

$$9 = x$$

Correct Answer : C

20. Solution:

$$\frac{1}{6} < 2x - 3 < \frac{3}{2}$$

$$\frac{1}{6} < \frac{6(2x-3)}{6.1} < \frac{3.3}{2.3}$$

$$\frac{1}{6} < \frac{12x - 18}{6} < \frac{9}{6}$$

$$1 < 12x - 18 < 9$$

$$1 + 18 < 12x < 9 + 18$$

$$19 < 12x < 27$$

$$\frac{19}{12} < x < \frac{27}{12}$$

$1.58 < x < 2.25$, since x is an integer, x can here a maximum value of 2.

Correct Answer : C

ANSWERS KEYS

Pretest Answer Key

1) C	2) D	3) B	4) C	5) B	6) B	7) B	8) C	9) D	10) B	11) D	12) B	13) B	14) A	15) C
16) D	17) D	18) B	19) C	20) B	21) D	22) A	23) C	24) B	25) C	26) B	27) B	28) D	29) B	30) A
31) C	32) D	33) B	34) A	35) C	36) B	37) C	38) A	39) D	40) B	41) B	42) C	43) D	44) A	45) B
46) D	47) D	48) D	49) A	50) C										

Order of Operations Answer Key

1	2	3	4	5	6	7	8	9	10	11	12	13	14	15	16
25	27	15	82	19	21	−11	15	27	3	1	$\frac{55}{3}$	−2	12	15	77

Fractions and Operations with Fractions Answer Key

1	2	3	4	5	6	7	8	9	10	11	12	13	14	15	16
$\frac{-2}{15}$	$\frac{1}{6}$	$\frac{1}{35}$	$\frac{13}{42}$	$\frac{43}{12}$	$\frac{-7}{10}$	$\frac{1}{5}$	$\frac{25}{72}$	$\frac{2}{81}$	$\frac{3}{4}$	4	36	$\frac{1}{27}$	$\frac{9}{2}$	$\frac{97}{12}$	$\frac{5}{6}$

Integers and Operations with Integers Answer Key

1	2	3	4	5	6	7	8	9	10	11	12	13	14	15	16
12	−32	$\frac{-8}{3}$	−50	$\frac{-2}{3}$	$\frac{3}{2}$	−1	$\frac{-1}{7}$	−60	$\frac{-1}{4}$	b / a	$\frac{-5}{3}$	$\frac{4}{5}$	$\frac{-5}{9}$	−40	24

AMERICAN MATH
ACADEMY

ANSWERS KEYS

Exponents and Law of Exponents Answer Key

1	2	3	4	5	6	7	8	9	10	11	12	13	14	15	16
2^{15}	1	6	$\dfrac{1}{5}$	1	$8x^3y^9$	$\dfrac{y^{10}}{9x^2}$	$\dfrac{9}{x^6}$	3^{18}	$\dfrac{3y}{x^2}$	$\dfrac{1}{3}x^2y^2$	$\dfrac{x^4}{y}$	y^6	2^{3x}	3^{x+1}	$\dfrac{1}{2^{5x}}$

Absolute Value and Inequalities Answer Key

1	2	3	4	5	6	7	8	9	10	11	12	13	14	15	16
$-1, 5$	No Answer	4	1	$\dfrac{3}{4}$	$x > \dfrac{4}{3}$	$x < 11$ $-1 < x$	$x > 8$ $x < -6$	$3 < x$	$x > -9$	$x < -13$	$\dfrac{18}{17} < x$	No.S	$x < 4$	$\dfrac{-5}{2}, \dfrac{5}{8}$	$x \pm 39$

Laws of Radicals Answer Key

1	2	3	4	5	6	7	8	9	10	11	12	13	14	15	16
12	3	$30\sqrt{2}$	$\dfrac{2}{3}$	$\dfrac{5}{3}$	$-3\sqrt{2}$	2000	$2\sqrt{15}$	$a\sqrt{a}$	$\dfrac{1}{x^2}$	7	$2\sqrt{3}$	$2\sqrt{a}$	$3\sqrt{5}$	$5\sqrt{3}$	$8\sqrt{3}$

Coordinate Plane Answer Key

A (4, 8) Quadrant I	B (–6, 6) Quadrant II	C (3, 3) Quadrant I	D (4, 0) No Q.	E (3, –6) Quadrant IV	F (0, –8) No. Q	G (–8, –5) Quadrant III

Factors & Multiples (GCF and LCM Answer Key

Prime Numbers	Composite Numbers	GCF (15, 45) = 15	GCF (16, 24) = 8	GCF (18, 48) = 6	GCF (5, 45) = 5	GCF (11, 88) = 11	GCF (5, 10, 15) = 5	GCF (16, 24, 36) = 4
2, 3, 5, 7, 11, 13, 17, 19, 23	6, 8, 9, 10, 12, 4, 15, 16, 18, 20, 21, 22, 24, 25	LCM (15, 45) = 45	LCM (16, 24) = 48	LCM (18, 48) = 144	LCM (5, 45) = 45	LCM (11, 88) = 88	LCM (5, 10, 15) = 30	LCM (16, 24, 36) = 144

ANSWERS KEYS

Scientific Notation Answer Key

1	2	3	4	5	6	7	8	9
1.25×10^7	1.23×10^3	9.8×10^{-7}	1.45×10^8	4.57×10^{-11}	8.69×10^{-11}	9670,000	0.00457	21,900

10	11	12	13	14	15	16	17	18
1.5	40,000	34,500,000	1.8×10^0	4.44×10^{-2}	3.784×10^0	8.9×10^7	3.67×10^0	2.34×10^5

Ratio, Proportions and Variations Answer Key

1	2	3	4	5	6	7	8	9	10	11	12	13	14	15	16	17
$\frac{2}{3}$	$\frac{8}{35}$	$\frac{1}{4}$	$1\frac{1}{11}$	No	Yes	Yes	No	$\frac{2}{3}$	$\frac{162}{7}$	-3	13	5	$\frac{3}{32}$	20	12	40

Unit Rate and Percent Answer Key

1	2	3	4	5	6	7	8	9	10
30	4	5	40	18	20	9	$20	$56.25	$30

Mixed Review Test I Answer Key

1) C	2) B	3) D	4) D	5) B	6) C	7) B	8) C	9) C	10) C	11) D	12) A
13) D	14) B	15) D	16) A	17) C	18) B	19) D	20) C	21) A	22) D	23) D	

Mixed Review Test II Answer Key

1) D	2) C	3) A	4) 2.4	5) B	6) B	7) C	8) D	9) C	10) C	11) A	12) C	13) B	14) D	15) A	16) C	17) A	18) B

ANSWERS KEYS

Solving 2-Steps Equations Answer Key

1	2	3	4	5	6	7	8	9	10	11	12	13	14	15	16	17	18
5	15	$\frac{-1}{4}$	2	5	4	45	−16	$\frac{127}{6}$	−112	−7	5	1	−49	$\frac{-44}{3}$	3	−60	−35

Solving Equations with Variable in Both Sides Answer Key

1	2	3	4	5	6	7	8	9	10	11	12	13	14	15	16	17	18
15	27	−1	−18	$\frac{15}{8}$	−14	84	0	−2	No.S	−3	$\frac{40}{3}$	−2	$\frac{2}{3}$	−2	0	0	12

Properties of Algebraic Equations and Simplifying Equations Answer Key

1	2	3	4	5	6	7	8	9	10
39	45	$ab + ac$	$4x + 12$	$-9x + 15$	$2x + 10$	$2x + 8$	$4x + 28$	$\frac{1}{2}x - 6$	$\frac{x^2}{3} - 2x$

11	12	13	14	15	16	17	18	19	20
$3ax + 6bx$	$2x - 3y$	$-3x + 6y - 12z$	$x + 2$	$10x^2 + 3x - 1$	$a^2 - b^2$	$x^2 - y^2$	$a^4 - b^4$	$\frac{1}{x^2} - x^2$	$a^4 - a^2b^2$

Solving Equations Involving Parallel and Perpendicular Lines Answer Key

1	2	3	4	5	6	7
Neither	Perpendicular	Perpendicular	Perpendicular	Parallel	Perpendicular	$y = 4x - 5$

8	9	10	11	12	13	14
$y = \frac{1}{2}x + 9$	$y = 5x + 1$	$y = \frac{2}{3}x + \frac{7}{9}$	$y = x - 5$	$y = -\frac{2}{3}x - \frac{14}{3}$	$y = x + 4$	$y = -2x - 4$

ANSWERS KEYS

Solving Systems of Equations by Substitution & Elimination Answer Key

1	2	3	4	5	6	7	8
$x = 9, y = 2$	$x = 9, y = \dfrac{7}{2}$	$x=-18, y=32$	$x = 3, y = 2$	$x=11, y=-10$	$x=22, y=36$	$x=-4, y=-3$	$x = 2, y = 8$
9	**10**	**11**	**12**	**13**	**14**	**15**	**16**
$x = 15, y = 5$	$x = \dfrac{9}{2}, y = 3$	$x=30, y=10$	$x = 24, y = 4$	$k = -1$	$x = -27$	$l = 7$	\$4.8

Factoring Quadratic Equations Answer Key

1	2	3	4	5	6	7	8
$(x - 7)(x + 7)$	$(x-10)(x+10)$	$(x-3)(x-5)$	$(x-7)(x+6)$	$(6x+3)(x-4)$	$\begin{array}{c}(x - 2\sqrt{2})\\(x + \sqrt{2})\end{array}$	$\dfrac{x^4 + y^4}{x^2 y^2}$	$(x+y)(x+y)$
9	**10**	**11**	**12**	**13**	**14**	**15**	
$(x+2)(x-8)$	$(x-8)(x-1)$	$(7x-8)(x+9)$	$(2x+3)(x-5)$	$(5x-1)(5x+2)$	Not factorable	$k=4$	

Solving Quadratic Equations by Formula and Complete Square Answer Key

1	2	3	4	5	6	7	8
$x = -2, -6$	$x = -1$	$x = -1, -\dfrac{1}{4}$	$x = 2, 6$	$\dfrac{1 \mp \sqrt{5}}{2}$	$x = 3, 1$	$3 \mp 2\sqrt{2}$	$\dfrac{-2 \mp \sqrt{13}}{2}$
9	**10**	**11**	**12**	**13**	**14**	**15**	**16**
$x = -7, 1$	$x = 2, -8$	$-6 \mp 2\sqrt{7}$	$3 \mp \sqrt{29}$	$x = 1, -5$	$-1 \mp \sqrt{7}$	$-2 \mp 2\sqrt{3}$	$x = -1, -3$

Adding and Subtracting Polynomials Answer Key

1	2	3	4	5	6	7	8
$5x^2+24x-9$	$-7x^2+18x-11$	$-x^2+17x$	$6x^2-7x-13$	$12x^2-24x-4$	$2x^2+9x-14$	$x^2+6x+24$	$-x^2-5x+13$
9	**10**	**11**	**12**	**13**	**14**	**15**	**16**
$2x^2+7x-9$	$-4x^2+5x$	$x^4-6x^3-17x^2+30x+60$	$6x^2+13x-4$	$-x^4-7x^3+17x^2-9x$	$-4x^2-11x$	$-6x^2-10x+13$	$4x^2+5x$

AMERICAN MATH
ACADEMY

ANSWERS KEYS

Multiplying and Dividing Polynomials Answer Key

1	2	3	4	5	6	7	8
$a^{13}b^9$	$a^{11}b^{17}$	$a^{18}b^3$	$6a^7b^8$	$a^{13}b^{23}$	$a^{21}b^{27}$	$2a^6-7a^5+3a^4$	$2a^2+6a+2$

9	10	11	12	13	14	15	16
$34a^{10}$	$3a^5-3a^3+21a^2$	$3a-5$	a^3+2a^2-10	a^3+5a^2+9	$6x^2-21x-12$	$\dfrac{a^2-b^2}{ab+a}$	$\dfrac{a-b}{a}$

Solving Equations with Algebraic Fractions Answer Key

1	2	3	4	5	6	7	8	9
$x=\dfrac{5}{2}$	$x=\dfrac{5}{3}$	$x=\dfrac{9}{7}$	$x=\dfrac{-13}{7}$	$x=\dfrac{56}{3}$	$x=384$	$x=\dfrac{-2}{25}$	$x=\dfrac{19}{2}$	$x=-20$

10	11	12	13	14	15	16	17	18
$x=42$	$x=6$	$x=\dfrac{1}{2}$	$x=4$	$x=-18$	$x=\dfrac{9}{5}$	$x=4$	$x=6$	$x=54$

Simplifying Rational Expressions Answer Key

1	2	3	4	5	6	7	8	9
$\dfrac{2x}{(x-2)(x+2)}$	$\dfrac{2}{-x-4}$	$\dfrac{3x^2-x}{(x-1)(x+1)}$	$\dfrac{x+1}{3(x+4)}$	$\dfrac{2x^2+x+5}{(x+3)(x-1)}$	$\dfrac{4x-1}{3x(x-1)}$	$\dfrac{1}{6x^4}$	$\dfrac{1}{3x^{12}}$	$x+6$

10	11	12	13	14	15	16	17	18
$\dfrac{1}{3x^{16}y^2}$	$\dfrac{x}{3}+1$	$\dfrac{z^6}{x^{11}y^7}$	$3x+3y$	$\dfrac{x^2}{12}$	$x-y$	$\dfrac{6}{x^6}$	$\dfrac{5}{3}$	$x-y$

ANSWERS KEYS

Function Notation and Inverse Function Answer Key

1	2	3	4	5
Yes	No	Yes	Yes	Yes

6	7	8	9	10
No	$f^{-1}(x) = x - 4$	$f^{-1}(x) = \dfrac{x+54}{2}$	$f^{-1}(x) = \sqrt{x+3}$	$f^{-1}(x) = x^2 + 6$

Function Operations Answer Key

1	2	3	4	5	6	7
$-x + 14$	$-5x + 4$	$x^2 + 5x + 33$	$3x^2 + 8x + 15$	$3x + 14$	$3x + 22$	$30x - 75$

8	9	10	11	12	13	14
$30x - 15$	1	48	328	1374	$x^4 - 3x^2 + 2x$	$\dfrac{x}{x+2}$

Mean, Median and Mode Answer Key

1	2	3	4
Mean: 6, Median: 6 Mode: 6, Range: 8	Mean: 5, Median: 6 Mode: 7, Range: 6	Mean: a · 32, Median: $\dfrac{7}{24}$ Mode: No Mode, Range: $\dfrac{3}{10}$	$y = 14$

5	6	7	8
$z = 28$	$k = 48.5$	32	Range

Slope and Slope Intercept Form Answer Key

1	2	3	4	5	6	7	8	9
$m = \dfrac{5}{4}$	$m = \dfrac{7}{4}$	$m = 2$	$m = \dfrac{-13}{4}$	$m = -3$	$m = \dfrac{1}{3}$	$m = -\dfrac{2}{9}$	$m = -\dfrac{1}{16}$	$y = 3x - 19$

10	11	12	13	14	15	16	17	18
$y = 6x + 9$	$y = 5x + 6$	$y = -2x - 14$	$y = -\dfrac{1x}{3} + 6$	$y = -\dfrac{2x}{5} + \dfrac{1}{5}$	$m = \dfrac{1}{4}$	$m = \dfrac{-1}{5}$	$m = -7$	$m = 8$

AMERICAN MATH
ACADEMY

ANSWERS KEYS

Distance and Midpoint Answer Key

1	2	3	4	5	6	7	8	9
$d=5$	$d=\sqrt{137}$	$d=\sqrt{145}$	$d=\dfrac{1}{6}$	$d=6$	$d=\dfrac{3}{4}$	$d=\sqrt{17}$	$d=10$	$d=\sqrt{170}$

10	11	12	13	14	15	16	17	18
$d=\sqrt{173}$	$M=(1,7)$	$M=(-5,-2)$	$M=(3,3)$	$M=\left(\dfrac{1}{3},9\right)$	$M=(0,4)$	$M=\left(\dfrac{1}{8},0\right)$	$M=(8,5)$	$M=(4,5)$

Pythagorean Theorem Answer Key

1	2	3	4
$x=10$	$x=15$	$x=2\sqrt{10}$	$x=9$

5	6	7	8
$x=15$	$x=10, y=5$	$x=15$	$x=\sqrt{73}$

Mixed Review Test III Answer Key

1	2	3	4	5	6	7	8	9	10	11	12	13	14	15	16	17
B	C	B	C	A	D	D	C	A	C	A	D	C	A	D	D	B

Mixed Review Test IV Answer Key

1	2	3	4	5	6	7	8	9	10	11	12	13	14	15	16	17
D	B	A	B	D	A	C	D	C	D	A	C	D	D	C	B	D

Final Test Answer Key

1	2	3	4	5	6	7	8	9	10	11	12	13	14	15	16	17	18	19	20
A	A	D	D	C	C	A	C	A	C	C	B	D	D	D	D	B	A	C	C

Made in the USA
Middletown, DE
09 September 2024

60535208R00091